Artificial Intelligence Business

Artificial Intelligence a modern approach.
Data Science, Data analytics and ML applied
in Healthcare, Marketing for Real World.
Successful business tools in Practice

For people, leaders and students that want to understand how AI can improve and boost up businesses

****

Copyright © 2019 by Jeff Mc Frockman.

Table of Contents

INTRODUCTION 1

OVERVIEW 4

CHAPTER 1: WHAT IS ARTIFICIAL INTELLIGENCE IN BUSINESS? 7

CHAPTER 2: A MODERN APPROACH ON ARTIFICIAL INTELLIGENCE IN 21ST CENTURY 18

CHAPTER 3: OVERVIEW OF DIFFERENT PROGRAMMING LANGUAGES: MODERN AND OLD APPROACH OF AI TECHNIQUES.. 24

Chapter 4: Differentiating Artificial Intelligence & Machine Learning terms 31

CHAPTER 5: The Most Important Artificial Intelligent (AI) Systems... 37

CHAPTER 6: REGULATION & ETHICS OF ARTIFICIAL INTELLIGENCE COMPANIES: SAFETY, LAW, AND GENERAL DATA PROTECTION 44

Chapter 7: ARTIFICIAL INTELLIGENCE IMPACT IN BUSINESS ASPECTS: OPPORTUNITIES, BENEFITS AND RISKS 51

Chapter 8: Data Science and Data analytics in Artificial intelligence 61

CHAPTER 9: ARTIFICIAL INTELLIGENCE APPLICATIONS IN FUNDAMENTAL SECTORS 67

CHAPTER 10: Growth of business value from Artificial Intelligence 82

Chapter 11: Chatbots and autoresponders..............................91

CHAPTER 12: THE RIGHT ARTIFICIAL INTELLIGENCE APPLICATIONS FOR YOUR BUSINESS 108

CHAPTER 13: ARTIFICIAL INTELLIGENCE FOR MARKETING AND ADVERTISING 119

CONCLUSION 142

INTRODUCTION

Congratulations for purchasing your copy of "Artificial Intelligence Business - Data Science, Data analytics and ML applied in Healthcare, Marketing for Real World. Successful business tools in Practice a Modern Approach Handbook". I'm excited that you have chosen to immerse yourself in the interesting world of Artificial Intelligence to help you in your business expansion. You will soon make discovery that there are various Artificial Intelligence applications that are in use in the modern world these apps are designed to help in making our everyday simpler and more productive.

Artificial Intelligence refers to the capability of digital computers or robots that are computer-controlled to carry out tasks normally linked with beings with high intelligent levels. AI is commonly used in the development of systems that are gifted with intellectual processes that characterize human beings i.e. the ability to discover meaning, learn, or generalize from experiences from the past. Demonstrations have shown that AI programmed computers or machines are well capable of handling very complex tasks.

Artificial Intelligence can be referred to as a concept in which a product, a robot, or a computer is made to think in a similar or a better manner than human beings. Artificial Intelligence (AI) can be defined as the study of how the human brain thinks, learns, makes decisions and works, in the process of problem solving. The main goal of AI is in the improvement of computer functions directly related to the human knowledge i.e. problem-solving, learning and reasoning

The main goal of Artificial Intelligence even as it continues being incorporated in business is to have the ability to adapt to and learn from broader array of challenges. This will eventually eliminate the risk associated with limitation in regards to AI problem solving capabilities.

At the same time, it's important to remember that as much as there is the positive aspect regarding how Artificial Intelligence impacts business, AI also has some risks associated with it. These risks can be very detrimental in your business though many technocrats argue that the benefits outweigh the disadvantages.

Artificial Intelligence isn't targeted to act as a total replacement of human beings. AI only serves in augmenting our abilities as well as enabling us to perform our tasks in a better and more efficient manner. Machines that are fitted with AI applications learn in a different way as compared to human beings; this means that they analyse things in a different way. They are better placed to capture patterns and relationships that escape us.

Artificial intelligence does an incredible job in business as far as delivering accurate results is concerned. This is simply because machines aren't affected by external factors as they carry on their duties as is the normal occurrence in human beings.

The most fascinating aspect is the use of AI systems in the medical sector; techniques from object recognition, image classification and deep learning are now being employed to spot cancer with similar accuracy as radiologists (through MRIs). Generally, there are several benefits that are directly linked to Artificial Intelligence as well a number of risks associated with the same. The main impact in the business world is the aspect of time saving and simplifying complex tasks.

There are several books on Artificial Intelligence a Modern Approach, thanks once again for picking this one! Every effort has been made to ensure that it delivers as much information useful for your business as possible.

OVERVIEW

Artificial intelligence (AI), is a concept whose invention goes back to 1956. Thanks to improved computing storage and power, advanced algorithms, and amplified data volumes,
AI has become increasingly popular in the present day.
AI is best described as the human intellect processes simulation by machines. These processes are inclusive of learning, reasoning and self-correction. Specific Artificial Intelligence applications include machine vision, expert systems and speech recognition
Artificial Intelligence enables machines to take to new environments, learn from past experiences, and to mimic human beings in performing their tasks. Most AI models that are in place today i.e. driverless cars, greatly rely on language processing and deep learning. By making use of these applications, computers have been empowered to perform particular tasks by recognizing data patterns and processing huge data volumes.
Artificial Intelligence automates learning that is repetitive and discovery via data. AI is very dissimilar from robotic automation that is hardware driven. Instead of opting for the automation of manual tasks, Artificial Intelligence performs huge volumes of computerized tasks reliably and frequently without getting tired. This automation type requires human

inquiry in setting up systems and asking the appropriate questions.

Artificial intelligence is very important in business in that it enhances intelligence to the existing products. In other words, products that you are already using will be upgraded with AI skill. Smart machines, bots, conversational platforms and automation can be joined with huge data volumes; this combination improves several technologies in the business world i.e. from investment analysis to security intelligence.

Artificial Intelligence is a program that makes adaptations through learning algorithms progressively; this allows data to carry out the programming. AL looks for regularities and structure in data which enables skill acquisition for the algorithm. This algorithm then becomes a predictor and a classifier.

Artificial Intelligence applications come in handy in analysing deeper and larger data volumes; this is accomplished via neural networks which contain several hidden layers.

Artificial Intelligence is presently being used in various sectors in business including:

Health care

AI programs have the ability to provide x-ray readings and medicine prescriptions that are personalized. Personal health care helpers can actually step in as life coaches; you are promptly reminded to eat healthier, take your medicine or exercise.

Retail Sector

Al offers virtual shopping abilities that provide recommendations that are personalized and engage the consumers in discussions concerning available purchase options.

Other business sectors in which Artificial Intelligence plays important roles include; manufacturing and banking sectors.

Researchers have found out that Artificial Intelligence is one very vital component in the running of business in the modern world. Several organizations have incorporated AI in their systems since AI has several positive impacts in business. Certain concerns have been raised as far as Artificial Intelligence is concerned. However, it has been impossible to overlook the benefits of AI in businesses as a whole. Prior to installing AI systems in your business, take time to find out whether the pros and cons of AI will work for or against you.

CHAPTER 1:
WHAT IS ARTIFICIAL INTELLIGENCE IN BUSINESS?

Prior to making any observations with regards to the manner in which technologies related to Artificial Intelligence are impacting the business fraternity, it's of great importance to have a good understanding of the Artificial Intelligence terms. AI is a general and broad term which denotes to any form of software in computers which performs humanlike activities. These activities may involve easy to complicated roles including planning to more complex ones like problem solving.

Artificial Intelligence is defined as the process in which machines replicate the processes of human intelligence. Some of the processes associated with human intelligence include self-correction, reasoning (making use of rules in reaching conclusions that are definite} and learning (process of acquiring information as well as rules that govern making use of these information). The Al has precise applications including; expert systems, speech recognition and machine vision.

AL is considered to be either strong Artificial Intelligence or weak Artificial Intelligence. Weak AI can also be referred to as the narrow AI; this is a system aimed and designed for precise tasks i.e. Apple's Siri. Strong AI is also known as general intelligence, a system that comprises of indiscriminate human intellectual abilities. When a task that isn't familiar is presented, the strong AI is tasked with finding a solution; this doesn't require intervention from human beings.

Since staffing, software and hardware costs for Artificial Intelligence can be very costly, many dealers are adding AI modules in their typical offerings. AI is also made accessible in the service platforms; this means that companies and individuals are permitted to try out with Artificial Intelligence for certain multiple platforms and purposes prior to making any commitments.

Some of the popular Artificial Intelligence cloud are as follows:

Google AI

IBM Watson Assistant

Amazon AI

While the tools in Artificial Intelligence present a variety of functioning capabilities for many businesses, their use has raised ethical concerns. This is because these machines are programmed by human beings, which leads to the belief that there is bound to be a bias which needs close monitoring.

Some technocrats have come to the conclusion that Artificial Intelligence is a word that is linked closely to common culture; this has caused fears that are unrealistic in human beings concerning AI and the questionable expectations concerning how life and the work place will experience change. Machines will only react and act like human beings only if they are well equipped with adequate information in relation to the present world.

Artificial Intelligence is a component in computer science in which the development machines that are gifted react and perform tasks like humans is emphasized. Some of the tasks intended to be performed by computers that have been installed with A.I are:

A. problem solving
B. Planning

C. Learning

D. Speech recognition

E. Ability to move and manipulate objects

F. Perception

Artificial Intelligence can be termed as a computer science branch whose aim is creating intelligent machines. This particular branch has become very vital in the technology industry.

How is Artificial Intelligence Applied in Business?

Artificial intelligence has been known to be of great impact in the business sector for many years now; it's been in practice for decades. Nevertheless, due to the possibility of accessing huge amounts of data as well as increased processing speeds, Artificial Intelligence has also started taking root in our lives on a daily basis.

From image or voice recognition and language generation to driverless cars, machine learning and predictive analytics, AI systems are applied in various areas. The AI technologies are crucial in yielding innovation, reshaping the manner in which companies carry on with their operations and in the provision of new opportunities in business.

Artificial Intelligence is speedily becoming a tool that is being used competitively in business. It's very clear that companies are beyond negotiating about the advantages and disadvantages of Artificial Intelligence. From data analytics to customer service to arriving at predictive recommendations, AI is viewed by leaders in the business world as a very necessary tool.

Artificial Intelligence has been ranked among the top technologies that a company needs to utilize; this can be done by discovering how to make use of the AI technology to their advantage. There are several devices that we use in our everyday lives that are enabled by the Artificial Intelligence applications i.e. smart assistant devices or apps like Apple's Siri or Amazon's Alexa.

Evaluating How Business Leaders Make Use of Artificial Intelligence in Competing Against Each Other

A good idea is making use of the internet and so the relevant research concerning how other companies make use of AI technology for their advantage.

You can also read through your competitors' social media platforms and websites i.e. Facebook and LinkedIn. You can also browse their blogs, news coverage, and press releases. You can also look manually for annual reports or newsletters that may not be posted online.

You can cast the net wider and do a search i.e. how hotels using Artificial Intelligence, or even 'how companies are using Artificial Intelligence.

Researching on the manner in which additional components of your chain of supply and support make use of Artificial Intelligence. Don't forget to research about other avenues that are non-digital. If you're attending an event that is related to your industry, look out for Artificial Intelligence sessions. Go ahead and speak to people sitting or standing next to you. You could also read journals which will enable you to find out as much as you can about your competitors. Be sure to get information on how they make use of AI to their advantage. Always note down any useful information that can help you in the implementation of Artificial Intelligence in your business.

There are several types of Artificial Intelligence applications used in the business world but the most commonly applied one is the Machine Learning software. This particular application is basically tasked with processing huge data volumes, at a fast pace. This could also mean the reduction in the need for manual labour or even result in the over reliance of AI in performing various tasks in the business environment.

How to decide whether Artificial Intelligence will work for you in your Business

Several factors can act as your guideline when it comes to making the decision as to whether to make use of AI in your business. You need to have prior idea based on your findings concerning what AI has done for other companies. The companies you base your research on need to be in similar industry, and if possible, similar size.

Understanding that various tasks require certain data types to work will enable you to make better decisions on the AI system to acquire for your company. Make sure that you fully grasp the limits and requirements of the intended tasks.

Be sure to consider the types of process that can be carried out by Artificial Intelligence. This gives you an understanding of what AI will be utilized for, ensuring that the choice you make will yield positive results for your business.

Is Artificial Intelligence a Blessing or a Threat in Business?

Technology is a very vital component in the growth and development of humans. AI is among these very vital technologies which are gaining hype and momentum. Technocrats have argued that AI can be a disaster and for some, a blessing.

Some of the advantages of Artificial Intelligence in Business include:

Using Artificial Intelligence decreases chances of making errors

This is mainly due to the fact that decisions made by machines are greatly influenced by data that had been recorded previously hence the chances of any errors occurring is greatly reduced. This can be termed as a great achievement since problems solving can be handled without leaving room for errors, even for the complex ones.

Business establishments that are advanced prefer using digital assistants in the daily interactions with their viewers; this plays a key role in time saving. This also helps businesses in fulfilling the demands from their users without any delays. These systems have been programmed to deliver the best assistance possible to users.

Artificial Intelligence facilitates appropriate decision making

Machines have absolutely no emotions which makes it very easy for them to make decisions that are more efficient, within shorter time frames. This can be best outlined in the health sector. Integrating Artificial Intelligence in the healthcare facilities has greatly improved efficiency in the administration of treatments. This is through the minimization of the occurrences of inappropriate diagnosis.

Artificial Intelligence can be used in risky instances

There are some instances where safety for human beings becomes vulnerable. i.e. survival for human beings on ocean floors is very difficult. Machines that have been fixed with algorithms that are predefined are used in studying these ocean floors. This is a major limitation that Artificial Intelligence has assisted to overcome.

Machines have the capability of working continuously

Unlike human beings, machines don't suffer from fatigue, even after working for many hours. This is a huge advantage over humans; they require time to rest for them to work efficiently. However, the efficacy of machines isn't determined by external factors. This means that these external factors don't reduce the working efficiency of machines.

Some of the disadvantages of Artificial Intelligence in Business include:

Implementing Artificial Intelligence is costly

When totalling the repair, maintenance and installation costs of Artificial Intelligence, the proposition is clearly very expensive. This isn't viable for industries and businesses that don't have sufficient funds; it proves very difficult for the implementation of AI technology into the businesses' strategies or processes.

Using the Artificial Intelligence has greatly increased the reliance on machines

Human beings have become increased dependent on machines. In the coming times, it might get to a point where it will be impossible for human beings to perform their duties without the aid of machines. The dependency of human beings on machines is bound to increase in future. This will result in the cognitive capabilities of human beings decreasing with time.

Artificial Intelligence has resulted in the displacement of people from manual jobs

This is the main worry for technocrats. There is the possibility that Artificial Intelligence will end up displacing several low skilled workers. Machines have the ability to work non-stop which has resulted in companies preferring to use machines instead of using manual labour. The world is edging towards automation which will mainly result in the majority of jobs being carried out by machines i.e. in the scenario where there is the inception of the driverless cars, uncountable drivers have been pushed into unemployment.

Machines Fitted with Artificial Intelligence perform restricted tasks

These machines have been programmed to perform certain duties with regards to what they have been programmed and trained to do. Relying on these machines in instances that may require the adaptation to unfamiliar environments can be frustrating. Machines don't possess the ability to think in a creative manner since their reasoning is restricted around the specific areas they have been programmed to do.

In conclusion, rather than replacing human ingenuity and intelligence, Artificial Intelligence is generally viewed as a subsidiary tool.

Irrespective of the fact that Artificial Intelligence is finding it hard to complete obvious tasks presently, it's skilled in analysing and processing volumes of data quicker than the human brain. This is a good thing in relation to the business world in the sense that AI accelerates the rate at which tasks are carried out, saving time that can greatly be reflected in increased sales.

When it comes to making the decision about whether to make use of Artificial Intelligence applications in your business or not, always evaluate whether the AL will work for the good of the business or vice versa. Always make considerations about the cost implications as well as other factors like if the systems will have any impact in profits maximization. Will Artificial Intelligence help you in making informed decisions? Will the AI lead to rendering people who have worked for you for decades jobless? What impact will the same have in their livelihoods? These questions will enable you to make informed choices as you take the next step in installing AI systems in your business.

CHAPTER 2:

A MODERN APPROACH ON ARTIFICIAL INTELLIGENCE IN 21ST CENTURY

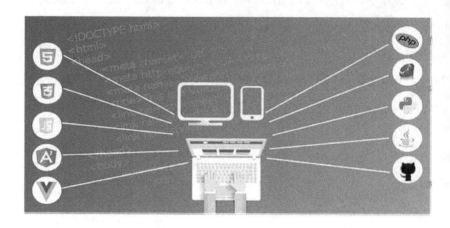

Learning is very fundamental, and particularly for the 21st century human being. Just when you think you have acquired enough, something else comes about, which poses a challenge for us to be on the move always. Essentially, when you discover new trends, your learning mode degrees change. Your level could move from surface to deep, depending on your interest and curiosity. Loosely speaking, as we seek to improve and get deeper into learning, the machines have also not been left behind. They are out to mimic human intelligence and they have proved to do even better than humans in some areas.

It has been many years of rule-based structures. Fortunately, with continued learning, techniques like deep learning have been discovered. Deep learning is one of the approaches to ML that have evolved quite swiftly. However, although it heavily relies on data and looks brilliant, the truth is that there are still numerous concerns that need to be addressed. We need to ask ourselves questions like 'how closely can technology imitate human intelligence?'

Exactly How Artificial Intelligence Resurgence in 21st Century Feels Like

Today's AI systems are not fully autonomous as yet. Even so, they are amazing as they can enhance the competence of human operators. The collaboration between artificial intelligence systems and the techies has brought about better experiences for both the customers and employees in the business setting. From the experts' desk, AI and its ML subset are on the verge of penetrating the military arena. There are proven military technologies that are falling into place, and they offer nothing short of an epic future for the industry.

AI as a modern approach to military technology

Artificial intelligence is one of the primary components predicted to impact war-fighting technologies. You may have heard of the Third Offset Strategy, a technique that the U.S forces use to detect asymmetries between them and their potential opponents. Being in the 21st century, military innovation is under watch for what it will do to enhance defense advantages for U.S forces. The key to the success of the Third Offset lies within artificial intelligence and autonomy. Employing these two aspects in the battle network will perfectly fit into the framework. It only requires one to appreciate the fact that several techniques have to be integrated.

Thankfully, the security and intelligence systems are already in the process of studying such systems as robotics and missile defense for deployment when it's the right time. U.S forces are not the only ones delving into military AI. Their near-peer opponents and competitors are also out to find a way of gaining a corresponding war-fighting advantage against them. Now, such competitive engagements will most likely result in dependence of inscrutable algorithms, which depend on cloud-level computing to determine the fate of the army. As we ponder on this, the most important thing is to prioritize professional military education to wield the said technologies. This can be achieved through democratizing experimentation and augmenting collaboration.

Artificial Intelligence: the Insights you Need

With a lot of improvement being experienced in the AI space, expert insights are always burgeoning. This is an interesting field where remarkable innovations can be discovered by any devoted researcher. As you read, keep your mind open to allow room for independent thinking as there is still more to be done before we can boast about attaining absolute success. Meanwhile, we will thrash out some of the insights that you need for your knowledge.

Understanding the position of data lineage: this is a crucial aspect that maps out the data's course. As you study AI and its techniques in profundity, remember that lineage has significant implications on each of them. These include neural networks, ML, deep learning, and natural language processing. Frankly, it's impractical to have accurate data-feeding artificial intelligence without data lineage.

Artificial intelligence in manufacturing: manufacturers are getting immersed in data daily, especially with the proliferation of networks and sensors in the working environment. We must appreciate that even the apparently negligible improvements have major implications. Artificial intelligence naturally finds affinity with manufacturing because the industry is heavily data-reliant. The truth is that we are on the threshold of a momentum shift.

AI use in shunning fraud in insurance application: submitting your insurance policy applications online sounds like such a deal, right? It is not only convenient but also quick to research on the rates and settle the entire process in a matter of several computer clicks. Sadly, fraudsters are into it too. They perform some malicious tasks like opening policies for imaginary beneficiaries, opening and revoking policies to make additional benefits and quota as well as applicants fabricating information to lessen their premiums.

Now that they possess loads of data, insurers can employ AI to verify the authenticity of every applicant.

AI in aiding proper prescriptions and combating illicit drug addiction: it is heart breaking to learn that there are thousands of people who die from prescription overdose every year. According to a report by the HHS, deaths resulting from drug overdose are manifested across the whole demographic spectrum. Opioid overdose has been notorious, and the upward trajectory is quite worrying. The techies can employ data-driven techniques to help save lives. Integration of pattern recognition, ML, and anomaly detection will help analysts to make remarkable advancements like;

Detecting early addiction signs in patients.

Predicting and impeding drug trafficking by quickly detecting suspicious prescription and dispensation patterns.

Ensuring clued-up prescription choices are made. This can be achieved through coordinating treatment whereby patient and drug details are passed to the doctors and prescribers directly.

Artificial intelligence insights and applications are not limited to the above batons. There are other fields of influence like incorporation of chatbots in analytic applications, employing AI in banking, marketing, drug development, and many more. The good news is that even with evident gaps in the AI strategies and capabilities, the industry leaders point out a potential outburst of its adoption. Government leaders and technologists should step up and learn from experiences by their peers. This is the only way through which we will all set out on a successful path.

CHAPTER 3:
OVERVIEW OF DIFFERENT PROGRAMMING LANGUAGES: MODERN AND OLD APPROACH OF AI TECHNIQUES

Computer programming languages are formal languages comprising a compilation of instructions that generate a range of outputs. They are primarily used in implementation of algorithms. These languages provide the most ideal platform for programmers to develop applications, games, software, etc. Some of them can be integrated while others can be considered to be independent. We cannot deny that this is an indispensable point to start from even as we scale to greater heights of machine involvement in performing human tasks. Programming languages are divided into 3 main kinds defined below.

1. **Machine language:** this is a set of binary digits that a computer can read, interpret, and respond to directly. It is the only language understood by a computer and it can differ depending on the OS on the computer.

2. **Assembly language:** this is any low-level encoding language intended for a particular type of processor. An assembler can be used to convert assembly codes into machine codes.

3. **High-level language:** a programming language that allows a designer to write codes that are independent of a certain computer type.

Programming language is essentially the heart of software. Without it we cannot communicate with systems as they only understand machine code. While human beings only understand high-level languages, machines do not. For the two to communicate there has to be an intermediate. Now, there are many programming languages and we cannot exploit them all. We will only look at some of them that are considered to be in demand. Before then, here are the types of programming languages that you need for better understanding of the languages.

Procedural programming language: one that follows a collection of commands in order. They happen to be some of the common types used by programmers and script. This could be because they employ conditional statements, variables, and functions to design programs through which computers can perform computations and display desired outputs.

Object-oriented programming language (OOP): just as the name goes, this is a model where programs are organized around objects, or data, rather than logic and functions. An object is simply a data field with a unique behavior and attributes. For instance, a person described by unique features like address and name is a good example of an object. OOP is based on four principles; encapsulation, abstraction, inheritance, and polymorphism.

Functional programming language: this is a language meant to influence implementation of software construction by building pure functions. Rather than emphasizing on statement execution, this procedure focuses more on declarations and expressions.

Scripting programming language: a language designed to communicate and combine with other languages. This language is often used alongside Java, C++, and HTML because it is not a fully-fledged programming language. They can get started even with a tiny syntax.

Logic programming language: this is a language that allows programmers to make declarative statements, after which it then gives room for the machine to judge the consequences of those same statements.

An Outline of Some Programming Languages

Python: this is an object-oriented programming language that is built on robust and flexible semantics. It is used by a variety of developers and software engineers, where it is preferred for its ability to quickly integrate systems like glue or scripting language. It is easily read and simple to learn. Python was developed in late 1980s and released in 1991. NASA uses it as a standard scripting language in their integrated planning system. The disadvantage is that it is slower and therefore not scalable for relatively large projects.

Java: equally object-oriented, java is a high-level language that is quite perfect for web development. It's portable across other platforms and does not need memory management, community support, or multiple compilations. Its syntax can get really long now that everything must be an object. Below is a simple code that prints 'Hello World.'

```
public class Hello {
        public static void main(String []args) {
                System.out.println("Hello World");
        }
}
```

C language: this is a structure-based language used in development of low-level applications. C is very fast and other languages actually inherit syntax from it. Unfortunately, it lacks flexibility of the object-oriented paradigm and does not do anything for you. You have to explicitly build everything and it does not caution when you're about to make a mistake. Below is a simple code for the output 'Hello World.'

```
#include<stdio.h>

int main(void) {
    printf("Hello World\n");
    return 0;
}
```

C++: this is an object-oriented, universal language, which is also middle-level and an expansion of C. It is an ideal illustration of a hybrid language now that you can code it in C or C++ style. C++ is very fast and possesses modern features, which makes it generally safer than C. The disadvantage is that it is complex to learn and also bears some readability issues. Here is a simple code that outputs 'Hello World.'

```
#include <iostream>

int main()
{
    std::cout << "Hello, world!\n";
    return 0;
}
```

JavaScript: it is a client-side scripting code which operates within a user browser, after which it generates commands. Its commands are processed on a computer and not on a server. It has no relation to java despite its name. In fact, it is usually placed on an ASP or HTML file. JavaScript is employed in web design not only to impact key page components but to also enhance their dynamism. Some of the pages include; scrolling abilities, generating a calendar, and printing date and time.

Which language do you choose?

When it comes to the choice of a programming language, none is perfect. Your preference should be guided by the circumstances surrounding the project. Time is also a critical factor that should guide your choice. For instance, you may decide to develop a basic tower defense game. While you could make the same game in Assembly, it is more prudent to consider Java. Why? When using Java it would take you a month or two to get done, while the same project would take you about a year, not to mention the headaches. If you are in this to find a simple language, you may need to first understand several of them. From there you will be in a better position to make an informed choice even as we look forward to advancing into modern techniques like artificial intelligence and its subsets.

Chapter 4:
Differentiating Artificial Intelligence & Machine Learning terms

Machine Learning &Artificial Intelligence systems are among the most popular catchwords in the business world; in many instances, they are put into use interchangeably. There is plenty of confusion when it comes to understanding and differentiating the two. The most common questions include; what is Artificial Intelligence? What is Machine Learning? Are the two related in any way? In most businesses however, marketing tends to ignore their differences for sales and advertising.

Machines and Artificial Intelligence have become an integral part in everyday living. This doesn't mean that the two are well understood. If you're hoping to implement and make use of ML or AI in your business, it's of importance to find out which one you plan to make use of. Artificial Intelligence and Machine Learning are related, but this doesn't mean that they are similar. Choosing either AL or ML can be the determinant in moving your business to the next level.

Artificial Intelligence basically means that machines are equipped to carry out tasks in intelligent ways. These machines are programmed to do various tasks by adjusting to various situations.

Machine Learning can be viewed as an Artificial Intelligence branch; the main difference is that it is more precise than the conclusive concept. The ML concept is founded on the idea that machines that don't require constant human supervision are created. Artificial intelligence has two major subcategories namely applied as well as generalized AI. In this case, the applied AI is pretty more relevant in that it covers various programs including driverless cars in order to import stock trading programs. While on the other side, generalized AI is not that common in practice since it's complicated to create. This type of AI has therefore, the ability to handle various tasks in the same way human would. Machine learning has been explored due to specific breakthroughs in Artificial Intelligence.

Therefore, such breakthroughs are specific in assisting humans to handle their tasks pretty more sensibly. Another breakthrough that led to the invention of ML was the internet. The internet has since facilitated extensive storage of information. This has never happened in the past. Machines can also be used in viewing data amounts that have not been accessed over the years in many ways. This is because of the past limitations in storage. As such, the volumes of data formed in excess for human processing makes it easier to perform complex tasks in shorter frames.

The Machine Learning Concept Explained

Machine Learning is an Artificial Intelligence branch which is tasked with providing systems with the ability to improve and learn from experiences automatically; this is achieved without making use of explicit programing. The main focus of ML is the manufacture and improvement of computer programs. These programs can actually use the data they acquire for their own learning.

The learning process starts with data observation i.e. direct instruction, or experience, with a view to looking for data patterns which will facilitate better decision making in future. The main aim is enabling computers to learn automatically and alter actions accordingly, without any assistance or intervention from human beings.

In ML, algorithms procure their skill or knowledge through experience. Machine Learning is an application which is reliant on huge sets of data in reminding the data about finding common patterns.

Artificial Intelligence Defined

Learning in Artificial Intelligence is through acquisition of knowledge as well as exploring ways in which the knowledge can be applied. The goal of AI is to arrive at optimal solutions by increasing the rates of success. Basically, Artificial Intelligence can be defined as an application in which computers make attempts to emulate or mimic things done by humans, in a better manner.

What are the Main Differences between the Machine Learning and Artificial Intelligence Terms?

AI is an abbreviation for Artificial Intelligence; intelligence defines the ability to acquire and make use of knowledge. ML on the other hand stands for Machine Learning which basically refers to acquisition of skill or knowledge.

Artificial Intelligence doesn't aim at increasing accuracy but increasing the chances of success. Machine Learning aims at increasing accuracy, but it doesn't really put emphasis on success.

Artificial Intelligence is a program in co putters that performs smart tasks. The Machine Leaning is a simple concept in which the machine gets to learn from collected data.

The goal of Artificial Intelligence is stimulating natural intelligence to enable it to solve intricate problems. Machine Learning aims at learning from data in regards to certain tasks inn maximizing the performance of the machines.

Artificial Intelligence is basically tool for decision making whereas the Machine Learning system learns new things from previous data.

AI leads in the development of systems that mimics human beings in responding to certain circumstances. Machine Learning is basically tasked with the creation of algorithms that are used for the sole purpose of self-learning.

Artificial Intelligence works by arriving at optimal solutions whereas Machine Learning works by arriving at solutions, whether ideal or not.

AI yields wisdom or intelligence. Machine Learning results in knowledge.

In conclusion, Applications in Machine Learning have the ability to read text as figuring out whether the author is offering congratulations or making complaints. These apps have the ability to listen to pieces of music and to arrive at decisions with regards to if the someone will be sad or happy after listening to the music. There are many possibilities that are provided by Machine Learning apps in connection with neural networks.

Machine Learning makes use of the experiences it handles to source for the learned patterns. Artificial Intelligence makes use of the same experiences in the acquisition of skills or knowledge and how to use the knowledge in handling new environments.

Both ML and Ai come with various business applications that are very valuable. However, Machine Learning has become more absorbed in solving serious problems in several businesses. The basic concept in applying the Machine Learning app and Artificial Intelligence system in business has been shown to play key roles in improved overall performance. The major disadvantage however is that making use of AI and Ml in your business automatically translates to loss of employment for manual workers.

CHAPTER 5:

The Most Important Artificial Intelligent (AI) Systems

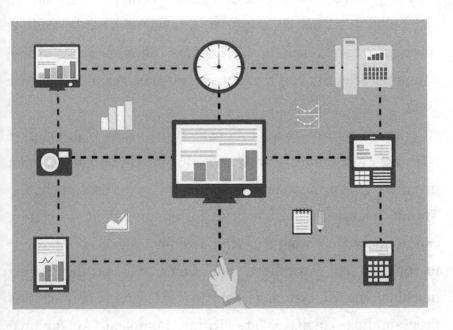

The ultimate aim of developing Artificial Intelligence systems is to replicate abilities possessed by human beings and operate just as intelligently as they do. To do so, developers approach each ability of human beings individually and develop efficient systems for each one. These human abilities include the cognitive skills to think and reason, perceive, learn, identify patterns, make decisions, act, solve problems, plan, remember, and use language skills. In this chapter, we'll look at how AI tries to perform these human abilities.

Artificial Intelligence Systems that act

Artificial Intelligence systems under this category act either humanly or rationally.

When AI systems perform intelligent operations indistinguishable from those of a human being, they are said to act humanly. This measure of intelligence is determined through a test called the Turing test, developed by English computer scientist and mathematician Alan Turin, in the year 1950.

AI systems are said to act rationally if their data manipulation rules sets can simulate what a Turing machine can do. Alan invented the Turing machine in the year 1936. Essentially, the Turing machine is the blueprint for computation. No system can do more than what a Turing machine can do. If, however, its operational capabilities can match that of the Turing machine, the AI system is said to be Turing Complete. All programming languages are Turing Complete. AI Rational systems rely on limited memory to analyze data and perform operations.

Rational agents, or bots as commonly known, are used to maximize the performance or efficiency of an operation. For example, a bot can perform task five or ten times faster than a Turing machine would.

Artificial Intelligence Systems that predict

AI systems that predict are an essential component of Artificial Intelligence for entities and are emerging as critical to the way they plan and set targets.

We use weather forecasts to make decisions on whether to dress lightly or heavily, depending on hot or cold weather forecasts respectively. Similarly, using their predictive models, entities such as businesses and governments use historical data gathered from their subjects to try and predict future outcomes. Every entity sets goals and targets of what they intend to achieve within a specified period. Using AI's predictive systems, they can get a more detailed foresight of what they will most likely achieve within that period. To make projections of future outcomes, the AI system analyzes relevant data from the subjects and puts all factors that impact on the entity's operations into consideration.

AI predictive systems are not entirely precisely accurate as yet, as AI is still a relatively new concept of computer science. With the future being impossible to forecast all the time authoritatively, this compounds the challenges with using AI's predictive systems. However, they try and give a good impression of what results will most likely be achieved during a specified period. With continuous development and improvement, AI predictive systems will get more and more accurate at predicting future trends and outcomes.

Artificial Intelligence Systems that learn

This form of Artificial Intelligence falls under the category of Generalized AI and is popularly known as Machine Learning.

Machine Learning aims to make machines independent and intelligent to learn from numerous kinds of unrelated data sets and apply reason and conclusions unique to each one. To perform this operation, unlike in rational agents, where the systems follow a specific set of instructions, Machine Learning systems rely on inference.

Machine Learning still requires massive development to achieve success as currently this form of Artificial Intelligence is riddled with many failures and is far from being proficient. With vast amounts of data and information available on the internet acting as fodder for Machine Learning systems to learn from, experts are upbeat that they will get better and employable. This increases the chances of success for the systems to produce desired results as purposed — to be independent, intelligent, and employable.

Machine Learning, therefore, requires a significant level of development to start being considered as a reliable commercial tool for entities. Nonetheless, this form of AI has a lot of potential for the future.

Artificial Intelligence Systems that create

Two Machine Learning systems are set against each another to make creations using AI successfully. One system creates an object, such an image, and the other works to find fault with the invention, thus making the two systems co-work autonomously. The concept of connecting two or more AI systems to work together to perform a task is what is known as Artificial Neural Networks, ANN.

An AI system that invents an object would first have to learn from similar data sets to replicate the creations, making its own. This operation is made possible using a new concept of Machine Learning called Generative Adversarial Network, GAN.

GAN has aided AI systems to make creations successfully. More recently, tech labs such as DataGrid and Samsung have successfully created both static and moving images using this method. Known as deep fake, AI systems can be able to generate very real-looking motion images from a single static image. Samsung AI Center was able to successfully use the famous portrait of Mona Lisa to make a video from it. Even more intriguingly, it was lip-synced to appear to speak.

As mind-blowing as this concept is, it poses a serious concern as people with bad intentions may use the technology for malicious purposes. Therefore, there needs to be tight regulation as to who can access this tool.

Artificial Intelligence Systems that relate

These systems learn the subject's location, habits, and preferences on their platform and relate the most relevant items the user is most likely to consume. The aim is to filter out unnecessary items and suggest the ones that would be most relatable with the user's data.

This tool of Artificial Intelligence is mostly used by search engines, social media sites, media content websites, and e-commerce sites to advertise products and suggest content relating to individual users on their platforms. It has proved to be an essential tool for driving up sales for e-commerce websites and increasing consumption within social media sites and media content websites.

Artificial Intelligence Systems that master

Developers aim to create a single computer system that can perform all intelligent tasks, unlike anything seen as yet in the Artificial Intelligence scene. It would arguably be an invaluable invention. Theoretically, it could combine all human beings' cognitive abilities already replicated in Artificial Intelligence functions and those still under development and perform all of them efficiently.

Given the superior power and abilities such a machine would possess, people have concerns and fears about it — and rightly so. Humans rule over creation because their abilities are superior to the rest of creation. If a single machine possesses all the skills of humans and more, then, these super intelligent machines would be superior to humans. They would have the power to rule over them, meaning either the extinction or insubordination of the human race. This chilling depiction is the picture of robots mostly portrayed in movies.

However, proponents of Master AI argue for the positives of these systems, such as having a single robot able to provide treatments to ailments and create a sustainable way of conserving the climate and enhance the security of citizens. Question is: "Can humans trust and tame a creation whose intelligence is significantly superior to theirs?" It would take a long time to find the answer to this question as the realization of this concept is considered presently unachievable.

Artificial Intelligence systems that evolve

In nature, the process of evolution aims at creating an improved form of a human being or organism to adapt better to its changing environment. Similarly, in the evolution of AI systems, a Machine Learning algorithm works in a topology of Artificial Neural Networks to create a more efficient and accurate system.

For over three decades, the most common method used in the evolution of Machine Learning algorithms is Gradient Descent, GD. GD identifies Neural Networks with the lowest cost functions. The lower the cost function of a system, the more efficient the system is at achieving the desired results. Therefore, Neural Networks learn from only the best algorithms and optimize to create a better one. To get best results projections, the number of Neural Networks has to be just right — not too few, not too many.

CHAPTER 6:
REGULATION & ETHICS OF ARTIFICIAL INTELLIGENCE COMPANIES: SAFETY, LAW, AND GENERAL DATA PROTECTION

Artificial intelligence has become a phenomenon in the present times because of the many benefits associated with it.In the business world, there are both benefits and risks that are associated with Artificial Intelligence. Businesses are making use of AI programs with the aim of improving their business results, profitability and productivity. The application of solutions driven by Machine Learning and Artificial Intelligence applications has become a very common practise. However, this doesn't mean that it is all smooth sailing for businesses using Artificial Intelligence.

There are several risks of artificial intelligence including:

Data Availability

During the phase of Ai implementation, data is often of low quality and inconsistent which makes it difficult to get value from AI.

Skills Shortage

There is the skills shortage problem for businesses trying to implement AI; there could be shortage of technical staff who are experienced. This could translate to extra costs to train employees for the successful operation of AI.

Implementing Artificial Intelligence is costly

Regular updating of software programs is needed to keep AI machines running smoothly. In cases where there is staff shortage, businesses are forced to outsource which is costly. The additional costs incurred in data training models can prove to be very costly especially for small businesses.

Other limitations related to Artificial Intelligence include:

There may be delays in the implementation process depending on the Artificial Intelligence programs being implemented

Misunderstanding the Artificial Intelligence programs which results in integration setbacks

Artificial Intelligence applications may not work well if interacted with other programs

These risks have necessitated the drafting and subsequent implementation of safety, law, general data protection regulation and ethics of artificial Intelligence companies. A framework tasked with the management of Artificial Intelligence ethics has been developed due to the significant risk potential associated with the same.

What are some of the core principles that guide governments, industries and developers in the ethical deployment of Artificial Intelligence driven systems?

Generating net benefits

It's a must for Artificial Intelligence systems to work by generating benefits for users which overshadow the risks.

Causing no harm

Artificial Intelligence must be designed in such a way that people won't be deceived or harmed and negative outcomes must also be minimised.

Legal and regulatory compliance

The systems installed with Artificial Intelligence must adhere to all government obligations, regulations and laws.

Protection of the users' privacy

Artificial Intelligence systems are compelled to see to it that data is kept confidential and protected. Harmful breaches in data must be prevented when using the AI systems.

Fairness

Individuals, businesses or communities using Artificial Intelligence systems must not be subjected to unfair discrimination. These systems should be implemented in a manner that they are won't cause unfairness which results from training biases.

Explainability and Transparency

Users must be well informed about how the AI algorithms that are being used affects them. Users must have proper understanding about the data or information that these algorithms use in making decisions.

Contestability

A competent process that can challenge the output or usage of algorithms must be in place since these affects users directly.

Accountability

Organisations and the people responsible for the development and implementation of the Artificial Intelligence algorithms must be accountable and identifiable for the resulting algorithm impact. This applies irrespective of whether the said impacts are intended or unintended.

How does the General Data Protection Regulation (GDPR) Safeguard Users in the Artificial Intelligence Centred World?

1) GDPR in privacy protection

The General Data Protection Regulation causes a hurdle in the implementation of Artificial Intelligence. At the same time, it is an important factor when it comes to the privacy of the users. Trust with regards to data privacy has been eroded as a result of data breaches. The good thing about this risk in Artificial Intelligence is that businesses are in a position to restore the lost trust simply by observing GDPR. This means that the companies take a straight forward approach that focuses in Artificial Intelligence privacy.

GDPR demands that holding data for longer periods than necessary isn't acceptable. This raises questions as to whether Artificial Intelligence is being silenced. This can be quite confusing since AI relies heavily on past data to process information that enables the system to make decisions.

2) GDPR in automated profiling and decision making

There are specific provisions of GDPR that target decisions that are based on Artificial Intelligence, specifically with regards to automated profiling and decision making. Set provisions must be adhered to strictly to ensure that the users do not get biased.

3) GDRP in information gathering

Ensuring that the company clearly understands the data that is collected and the manner in which the data is processed is the first as far GDPR is concerned. Documentation of the data type collected, the source and the channels via which it has been collected is crucial.

4) GDRP in risk assessment

Prior to making use of automated decision making, a Data Protection Impact Assessment (DPIA) must be carried out. DPIA allows the organization to gauge the risks that come with automated decision making. DPIA aims at examining risks to subjects of data at every data processing step.

5) GDRP in the management of third parties

It's important to have an understanding of the privacy and security controls that are used by your vendor. Be sure to find out from your vendor whether he has all the required industry certifications; this is very important in the verification of vendor assessments.

In conclusion, Artificial Intelligence and General Data Protection Regulation emphasizes that the big data that is used in Artificial Intelligence is directly opposed to purposeful limitation of data retention and collection. It's quite clear that as far the safety for AI users is concerned; following laid out regulations helps in safeguarding privacy of the users.

Be sure to find out prior to AI implementation that your vendors observe these guidelines to the letter. This may seem like a daunting task but the good thing about it is that users feel safer and more confident in transacting in your company when they have the guarantee that they are safe.

Chapter 7:

ARTIFICIAL INTELLIGENCE IMPACT IN BUSINESS ASPECTS: OPPORTUNITIES, BENEFITS AND RISKS

We come across the term Artificial Intelligence from time to time as we get along with our daily lives. There's more to Artificial Intelligence than 'Alexa' or 'Siri' which are great influencers in making our activities in the business world easier. The target of Artificial Intelligence is making computers to be smart enough to mimic the behaviour of the human mind. This basically means that Artificial Intelligence greatly helps us to achieve our business goals.

In the present age, Artificial Intelligence applications are in use constantly. You may have used the Ok Google feature in many instances but probably never gave it much thought. This Google feature is an application that make everyday living easier for human beings.

Artificial Intelligence in business is a concept that is very interesting and fascinating. Technology in the present day is picking at a very fast pace meaning that businesses have to keep up with this pace to ensure that they utilize technological advancements fully to their advantage. It's important to note that AI is being used in the improvement and transformation of businesses.

The Artificial Intelligence field is rapidly advancing. The AL applications enabled major breakthroughs in game playing, autonomous robotics, speech and image recognition. Mostly, AI promises many benefits; better and cheaper services and goods, medical advances and fresh scientific discoveries. At the same time, AI raises come concerns; security and safety, inequality, bias and privacy among others.

 Most of the AI programs in use today are very narrow; meaning that in some instances they may be incapacitated when it comes to handling certain tasks. The AI applications aren't necessary superior to human beings in all aspects; they could be having better performance than humans in one aspect or another but not all.

Where are Artificial Intelligent applications applied in Business?

The applications in AI are presently in use in different areas in the business field including:

- Business and sales forecasting
- Voice-text features
- Spam filters
- Security surveillance
- Automated insights
- Online support for clients over the internet and automated responders
- Process automation

It's very clear that the power that has been bestowed upon machines has helped greatly in solving business problems, but, at the same time it has led in destruction or risks in some instances. The following are some of the benefits/ opportunities and risks of using Artificial Intelligence in business:

Advantages of Artificial Intelligence in business discussed:
Reduction of the rate at which errors are made in business

Decision making by machines that are installed with Artificial Intelligence applications are purely base on prior data records and algorithms, greatly reducing the chances of errors occurring. This is a great accomplishment, as complex problems that need to be solved in businesses is done without any room for error.

Top business organizations are making use of digital assistants in interacting with customers. This accelerates the rate at which feedback is given to clients, providing the best quality assistance to users.

Making informed decisions

Artificial Intelligence applications have no emotions hence making them better placed to make unbiased decisions, in a short period of time. A good example is in the healthcare facilities; AI programs have facilitated provision of efficient treatments mainly by bringing down the risk of misdiagnosis.

Ability to perform tasks continuously

Machines installed with Artificial Intelligence programs can work for long periods of time without getting tired. This is a huge advantage in business as compared to human beings in regards to productivity. External factors never affect the efficacy of machines which is a very important benefit in business.

Disadvantages of Artificial Intelligence in business discussed:

Artificial Intelligence is very costly to implement and use

The total cost of installing, repairing and keeping AI machines in good shape is very high. Businesses with huge savings can implement the AI programs. However, companies that don't have sufficient funds will not be in a position to make use of these applications in their strategies and processes. This means that these business get locked out from enjoying the Artificial Intelligence benefits.

Displacement of manual workers

This is the main disadvantage of Artificial Intelligence; these apps have led to people being displaced from their places of work. This happens where a single machine gets incorporated in a business and performs a task that was being handled by several people, at a go. This automatically means that some employees are rendered jobless.

Over reliance on Artificial Intelligence Machines

Human beings are becoming increasingly dependent on AI machines; it's slowly becoming very hard for human beings to accomplish several tasks without the help of machines. This has led in decreased thinking and mental capabilities in humans.

Machines using Artificial Intelligence can only perform specified tasks

AI machines have already been programmed to perform only outlined tasks. These machines can't be relied upon to take to unfamiliar environments or make decisions outside the areas that they are programmed to handle.

How will Artificial Intelligence Impact Businesses in Future?

The future in the business world will greatly be determined by the use of various Apps in the simplification of various tasks. The Artificial Intelligence is most definitely among these Apps that are bound to play a key role in business running in future.

Machine Learning refers to the art of applying and designing certain algorithms that have the ability of learning events based on past occurrences. For example, complicated algorithms have the ability to analyse past cases of frauds and managing similar happenings beforehand. The only hitch in Machine Learning is that inn instances where there aren't any previous cases, then these algorithms can't work. Always remember that ML and AL have a number of differences but always work better when incorporated together.

By making use of Machine Learning and Artificial Intelligence applications in your business, the systems won't have limitations concerning the instructions that had been given out by the programmer. This means that the programs will not require plenty of programming compared to ancient programs.

The Artificial Intelligence app is bound to alter the manner in which businesses will be running in the coming days. Some of the changes that you can expect in future in business as a result of using Artificial Intelligence include:

Finding better candidates in a quicker and easier manner for your business.
In the work place environment, we depend a lot upon machines to do or complete certain tasks for us. Artificial Intelligence will be used in the identification of certain factors i.e. the period of time an employee is positioned in a specific place, the cost of hiring an employee, specific job positions that tend to be filled more quickly as opposed to positions which take much longer to be filled, etc. This will help you in understanding the behavioural trends of your employees.

There may be the increased ability in the identification of any hindrances while carrying out the recruitment process; this can even assist you in cost reduction in this process. This also ensures that you get to hire the most suitable candidates in the shortest time possible.

Saying goodbye to theft and errors

It's very normal for human beings to make mistakes. No matter how minute these errors appear to be, they end up affecting your business in one way or another. There is a great concern that in the coming days, machines or robots with Artificial Intelligence programs will eventually render people jobless.

These machines are quickly becoming popular in businesses since they increase business efficiency greatly.

If the AI system is developed properly, then human errors can be done away with completely, creating an environment that is almost risk-free for the business.

Companies that depend heavily on smartphones, laptops and the internet are always prone to theft or cyber-attack. Once AI familiarizes itself with the decodes and system deviations, algorithms and patterns, it reveals malfunctions and attacks in process, which prevents the business from incurring possible massive losses.

Simplifying and Upgrading the customer service process

This is among the most critical aspect in any business. It can be a task that is also very time consuming. You can imagine taking several hours to solve a single complaint from a customer. But with the trend in which customers raise their complaints via chatbots, the time consumed in resolving the same is really reduced.

Although there is no personal touch when using chatbots, with time, they will. Tools such as the Digital Genius have been implemented in creating human-like conversations that are flawless with customers. With the aid of Artificial Intelligence programs, companies are better placed to increase their data collection capacity and efficacy. This data is then processed which helps business owners to know the behaviour patterns of he customers. This will be vital in updating the policies of the business.

Measurement of Big Data

AI and Big Data are terms that will be key players in the future for the successful running of businesses. Artificial Intelligence businesses will not be in a position to work in the absence of Big Data in future.

It's important for business owners to ensure proper data preservation in vast amounts, which will ensure that no information is lost. Machines that have the capability of carrying out content analysis and reasoning that is evidence-based can improve the making of decisions for businesses edged towards people's management.

Branches of Artificial Intelligence I.e. Deep Learning and Machine Learning can be used in the management and analyzation of big data. It's basically used in sourcing for predictions, patterns and trends. This will be of great impact and importance in a finance business or stock market.

Lead generation;

This is defined as a process involving the cultivation and identification of potential customers for any business services or products. Expenses that are directly linked to the acquisition of customers through lead generation are considered to be very important. Artificial Intelligence businesses are making use of certain AI programs that enable better data processing as compared to human beings. These programs analyse data from social media platforms which enables business owners to have a better understanding of the interests of their customers.

In conclusion, the above mentioned business areas can be grasped better with assistance from Artificial Intelligence programs. Artificial Intelligence also assists in several areas in the business world such as time management and administration and many others. This shows clearly that Artificial Intelligence is undoubtedly going to become the key feature in business, in future. It would be a great idea to adapt and start making use of Artificial Intelligence for better and greater success.

Chapter 8:
Data Science and Data analytics in Artificial intelligence

Nowadays, it is utterly impossible not to come across analytics and data science mentions in modern online or physical environments. In short, analytics and data science or combinations of the duo are some of the terms that are trending in our present world. Consequently, various foundations now offer jobs based on full conversance to data science and analytics. Therefore, it is an understatement not to privilege the substantial data sciences and analytics' 'tech' invasion in corporate and academic environments after the swift evolution of the internet and digital economies. The intelligent Data

Science and Analytics technologies are gaining adaptation in industries, individuals and business establishments.

Data Science and Analytics involves the gathering, grouping and examination of data and connote its outlines, facts and aptitude. Concurrently, Data Science and Analytics can be classified on the base of their descriptive, predictive and prescriptive aspects. However, these aspects need to be theoretically, technologically and methodically developed to facilitate the growing global demands.

Innovation in the Data Science and Analytics Environment

Although certain basic constituents of Data Science and Analytics have long been in existence, their presence has overseen tremendous, innovative and opportunistic advancement in Machine Intelligence. Analytics as services are relatively new additions to the Artificial Intelligence world. In the accord of the involvement of analytical systems in data comprehensions, rapid migration of fiscal and social deals in the online world has been perceived. This has, in turn, facilitated the digital apprehension of vast statistics. Moreover, there has been a significant development in the contemplation of human exposition configurations and expositions. Consequently, these developments have stretched out the scope and accessibility of statistic sets. Accordingly, an exponential boom of inquiry expositions extending to various ranges of investigations has been observed. Also, boosted by the innovation of 'smarter' machines, the nature of inquiry has also improved. The new 'smart' machine features have been brought by, by the elevation and concurrent instrumentation of algorithms into inherent interactions with the algorithms. In addition to the collection of data for exclusive human analysis or vital record-keeping, a more reasonable capture of data backlogs, entirely, for un-envisioned hypotheses has been introduced. With the characteristic of optimal gather and understanding of data, computers have evolved to 'intelligent' probing machines. Computers, through refined probing, can consequently create and discover new knowledge that would be, long due in reliance to human hypotheses (Agarwal and Dhar,). This proves and justifies a probability of theories generated by computers.

However, with the evolution of algorithms come elevated probing desires. These desires are key building blocks in the improvement of algorithms and structures. Consequent to the increased, saturated and unstructured data, further evolution of algorithms is evident further focused on heterogeneous and uneven data. The research frontier is now involved in the creation of cutting edge developments in the areas of such as; transcript and image processing, and natural language processing, consequent to the uncertain existence of any or ultimate data incorporation and analysis limits transversely present on the internet.

In support of existing networks concerning humans and product ranges, a requirement of better sampling and validating algorithms are needed in the discourse of these 'societal' features (Ugander et al. 2013). These new developments involving interactive and encompassed algorithms evolve from a range of subjects including computer science and IS or any other disciplines. In concurrence to Kramer et al. (2014), research proves the manipulation of internet users' moods through social-based algorithms. Casual networks can be derived from variables through experimentation of people. However, conducting these experiments without informed approval may implicate the research with unethical privacy conduct. Nevertheless, proof of medium for the development of theories based on the investigation of social and other disciplines is evident.

Data science and Data analytics on Reasonable Benefit for Information Systems

As accorded by Aral and Walker (2009), this is arguably a golden era for Information Systems research. As evident from information incorporated in some commonly visited scientific channels such as PNAS, Data Science and Data Analytics research is routed in Information Systems (Aral and Walker, 2012). Consequently, common questions of the reasons for our choices and reasons for interest in Data Science and Analytics have risen. What influences the choices we make in social networks... sway or homophily? Accordingly, this and other questions offer scientists opportunities to discern, inquire, and connote social behaviors on an inclusive scale.

Data science and Data analytics on Investigation Questions

As coined by Lucas et al. (2013), following the Information Technology revolution, Data science, and Data analytics sanction the creation and implementation of studies. Moreover, they facilitate extends to existing research under explored outlines such as; learning result mediations, and satisfaction, online tutoring, and additional information of exceptional gauges. Furthermore, Data science and Data analytics facilitate the decoding of a network; distinct, organizational, and societal existences and their connoted outcomes.

Data science and Data analytics, Predictions and Explanations

As proven by Weil (2014), it is justifiable to acknowledge the importance of Data Science and Analytics in the providence of remarkable randomization of research from unrealistically massive observational data. These operations are applicable if data and aspects of data can be satisfyingly and respectively explained or predicted. In this accord, Data Science and Analytics offer researchers potentials for innovative discoveries.

Through Data Science and Analytics, scholars, researchers, and everybody have the advantage of creating new scientific prospects. As coined in the above Data Science and Analytics integrations, it is a superb and exciting period for research to evolve beyond Information Systems and general 'science'.

Concurrently, the delivery of data in the most accurate and easy forms are possible. In these accords, growth in education, technology, societies, businesses and other aspects of human existence is inevitable. Data science and analytics offer managerial prepositions to businesses since they are integrated with technological systems i.e. Artificial Intelligence and thus help improve decisions in businesses. With Artificial Intelligence abilities to plan, organize, lead and control, the management of the businesses is made easy through innovation, prediction, and investigation.

CHAPTER 9:
ARTIFICIAL INTELLIGENCE APPLICATIONS IN FUNDAMENTAL SECTORS

No doubt artificial intelligence is at the edge of transforming every section of our economy. It is exciting to see the many applications that AI accommodates in today's society, and more so the significant changes that it is making in those sectors. Allowing computers to make human-like decisions helps a lot in automation of routine tasks in businesses. The system learns from experience such that it is able to understand every customer, examine their behaviour, personalize experiences, and significantly cut back on operational costs. In fact, artificial intelligence is likely to become an indispensable tool in all medical disciplines. As we seek to understand how best AI disrupts and shapes different industries, we will first look at its application in the healthcare sector.

Healthcare and Medical Diagnosis

As AI continues to infiltrate diverse aspects of our lives, the impact is quite perceptible in the healthcare industry. Although there is already a lot to show for it, there is still wide-reaching potential in artificial intelligence in healthcare. Do you know that drug discovery and mobile coaching remedies are under the umbrella of things being achieved through machine learning? Well, let us get to the details on the much that we know AI has done in this industry.

Faster development of drugs

There are AI-powered firms that are set to collaborate with giants in the pharmaceutical industry so that they can significantly augment the efficacy of drug development. Undoubtedly, artificial intelligence has remarkable computational power considering that it employs personified knowledge gained from solutions. Honestly speaking, the pharmaceutical industry is experiencing major challenges in upholding their drug manufacturing programs.

Reduced efficiency and increased costs of research and development are just some of the major setbacks making the entire process notoriously expensive. The effectiveness of most of the analytical practices in drug development can be enhanced through AI and machine learning. Drug development incorporates four main stages which include; identifying target molecules for intervention, discovering effective drugs, expediting clinical trials, and finding biomarkers for diagnostics. Amazingly, artificial intelligence has already been employed in the four mentioned stages and was successful.

Identification of target molecules for intervention: this initial stage is seeks to understand the pathways, or the biological source of a disease and also its resistance ability. From there, you then have to discover positive targets for combating the disease (usually proteins). Discovery of viable target molecules requires a lot of data. Thankfully, the extensive availability of high throughput systems such as deep sequencing has made it easy to access more data. Even so, integration of an array of data sources with the traditional methods remains a challenge, and this is where artificial intelligence now comes in.

Discovery of effective drug candidates: upon identifying a target molecule, you need to now get a compound with the ability to interact with it (the identified molecule). The process entails screening of thousands or millions of probable compounds to gauge their affinity as well as toxicity. The compounds in question could be artificial, natural, or bioengineered. Unfortunately, the software in use at present is unreliable as it is frequently erroneous and has lots of false positives. As such, it always takes unusually long to identify the optimal drug candidates, usually known as leads. Now, with incorporation of AI, it is easier to have the ML algorithms learn how to forecast the correctness of a molecule. This can be achieved by incorporating molecular descriptors and structural fingerprints, a process that saves a lot of time.

Accelerating clinical trials: the process of finding appropriate candidates for experimental trials is quite difficult. Unfortunately, selection of the incorrect candidates ends up lengthening the trial, which eventually costs a lot of resources and time. Machine learning can effortlessly expedite the clinical trials design. In this case, ML can be used to automatically identify fitting candidates. At the same time, the ML algorithms can help in ensuring proper distribution for clusters of trial participants. The best thing about algorithms is that they can detect an unproductive clinical trial from the onset and send an early warning, giving room for the researchers to chip in and save the development process.

Find biomarkers for diagnostics: it is only possible to offer treatment to patients when you are guaranteed of their diagnosis. Some of the techniques used to diagnose the diseases do not only require high expertise, but are also very expensive. Whole genome sequencing is one such method that also requires use of sophisticated lab equipment. Biomarkers are molecules contained in human blood. These are the particles that ascertain whether or not a patient suffers from a given disease. Biomarkers, therefore, are quite effective when it comes to diagnosis, making the entire process cheap and secure.

These molecules can also be used to know the progression of a disease, which makes it easy for the medics to monitor the effectiveness of the drug. But again, it is also very expensive to determine the appropriate biomarkers for a particular disease. You'd have to screen hundreds of thousands of probable molecule candidates. At this point, AI can be used to automate a significant segment of manual work and classify the bad and good candidates, hence aiding the doctors in focusing on analysis of the best prospects. Biomarkers can be used in; detecting a disease at the earliest possible stage, identification of the risk of someone developing a disease, any likelihood on progression of some disease, as well as monitoring the response of the drug on a patient.

Diagnosis of diseases

Correct diagnosis of ailments takes years of arduousness and time-consuming processes. Diagnostics in most disciplines usually demand a number of experts far exceeding the available supply. As a result, this strains the available doctors while at the same time it impedes life-saving diagnostics. Thankfully, ML, mainly deep learning, has made significant advances in automatic diagnostics, making it economical and more accessible. Machine learning algorithms are smart such that they can learn to see patterns just like the doctors do. However, they need thousands of neatly digitized and concrete examples as they cannot read between the lines in study materials.

Please note that ML algorithms are particularly helpful in fields where the diagnostic data examined by a doctor is digitized. Such areas include, but are not limited to; detecting strokes or lung cancer by use of CT scans, evaluating the risk of cardiac arrest using electrocardiograms and cardiac magnetic resonance imaging (MRI), categorizing membrane lacerations in skin images, and identifying diabetes complications of the eye from the relevant images. The algorithms have become smart in diagnostics especially with availability of plentiful good data in the said cases. The good thing about them is that they are naturally data-heavy, and hence ideal for AI application. AI is not replacing the medics any time in the near future. It is more of a collaborative affair.

Personalize treatment

To increase the lifespan of a patient, treatment has to be tailored. As it is, different patients have different responses towards treatment schedules and drugs. However, identifying the particular factors that affect the treatment option is quite intricate. Machine learning has the capacity to automate this complex statistical task. The ML algorithms can learn how to effectively determine the characteristics depicting that a given patient will have a particular reaction towards some given medication. Precisely, there are algorithms that can help in projecting a patient's potential reaction towards a certain treatment. Usually, the system takes advantage such that it learns from cross-referencing comparable patients where it matches up their medications and results. Doctors rely on the resultant outcomes to design a reliable treatment plan.

AI for Real World:
Positive and Negative Aspects for Clients and Consumers

One undeniable fact is that AI-driven results have already had a massive impact on customer experience, and they will still continue to do so. It has reached a point where the clients and consumers acquire unprecedented information about a product before deciding whether or not to opt for it. Alternately, businesses have also benefited as they can use churn data to foretell the clients who're likely to leave or stay. Even with all the good, we cannot refute that there will come along some negative consequences as well. Discussed below are both sides of artificial intelligence for real world.

Positive Aspects of AI for Clients and Consumers

Customer satisfaction: this is the hallmark to delivering a first-class client experience. When coded properly, artificial intelligence has low fault rate as compared to humans. You realize that most client interactions require human involvement including social medial exchanges, phone calls, online chats, and even email. AI has made it possible for the computers to handle client enquiries with incredible accuracy, precision and speed. This is especially because when combined with ML, it becomes even better for the platforms to communicate with the customers, making their experience remarkable.

Simplified item search: if you have not known, then it is about time you appreciated that AI has already fortified its authority in the digital era. It is very likely that you interact with it on a daily basis, even though in its subtlest and most modest shape. In your phone or computer operations, you might have noted that there are ads that keep popping. Now, have you ever taken time to think how those ads always happen to be suited to your taste? Artificial intelligence monitors consumer behavior on social media and then goes on to influence their choice.

Intelligent customer service: it is very possible for artificial intelligence to gain essential client insight. This is where AI-powered computerized associates can learn a client's behavior and help them with selection by recommending products inclined towards their fit and needs. It builds a customer's confidence when they get a wide array of products that seem to match their needs and preferences.

Predicting outcomes: artificial intelligence is incredible in the sense that prediction of outcomes is possible by use of data analysis. For instance, when you have stock, the system can forecast the possibility of your commodities selling, and the margin by which it will happen. At the same time, it can also foretell when demand for your commodities will lower. As an entrepreneur, you need this fundamental information to determine the products to stock and the suitable amounts to source during a particular season.

Monitoring conversion rates: there are those traditional methods that people in the traditional days turned to for conversion. These techniques are no longer very effective especially for a modern business. When a potential client visits your site, you need to have a reliable way of turning them into customers. The said traditional method can even take several months, which means losing business. If implemented properly, AI can track all the essential customer records and draw a conclusion on the possibility of a visitor becoming a client. By so doing, the system helps you to save precious time that would otherwise be used to sleuth down inappropriate metric.

Negative Aspects of AI for Clients and Consumers

Job loss in some particular fields: while it is true that artificial intelligence will create many jobs, it is also likely that there are some that will be lost. There are those tasks performed by human beings that will soon be taken over by the machines. It means there will have to be a lot of training done so that our future workforce gets fully equipped. Of course there should be education programs that will need to be initiated to assist the current workforce to prepare for transition into new offices that will make use of their distinctive human abilities.

AI bias: remember that AI algorithms are created by human beings. As such, it is likely for them to contain in-built bias. This could be caused either inadvertently or intentionally by the persons introducing them into the system. Should the algorithms be bias or they get introduced to biased datasets, then it means they will automatically produce biased results. The results could cause unintended consequences just like it happened with the Twitter chatbot by Microsoft that depicted a lot of racism.

Accelerated hacking: even though there is great acceptance by the consumers and business owners to roll out AI-powered services, there is also the fear of privacy and data protection. After all, AI is good at expediting results in diverse instances, making it difficult for humans to follow along. That is why malicious acts like instilling viruses into software programs and phishing have become common.

Global regulations: this is very necessary for safe and effectual interactions globally. Regulation can only be done with introduction of artificial intelligence laws and regulations amongst governments. Since technology has made the world a global village, it means that we are not isolated from each any more. As such, the decision that a country makes regarding AI is likely to affect other countries, whether positively or negatively.

Artificial intelligence terrorism: technology is open for all and this means there could also be forms of terrorism depending on AI. Think of development of autonomous drones and robotic swarms. Our defense organizations will need to fine-tune all the potential threats. It will take extensive expertise and human reasoning to ensure minimal adverse impacts.

Artificial Intelligence for Banking: Critical Banking Functions Most Likely to be Transformed by AI

The financial industry, being data-weighty, has also proved that indeed artificial intelligence is able to change its landscape as well. Amazingly, this is happening even in the fields considered to be traditionally conservative. Without doubt, the banking sector is a fertile ground for artificial intelligence. If deployed properly in the next few years, AI will certainly boost revenues and potentially minimize costs. At the same time, there are high chances of a rapid increase in account and transactional security. This is especially with the expansion of cryptocurrency and blockchains. Looking at it critically, you also realize that when the intermediaries get reduced or eliminated, transaction fees will also be toned down.

With cognitive computing, all the applications and types of digital associates will continue to perfect themselves. Managing personal finances will only get exponentially easier as smart machines will do much of the work. Even tasks like organizing tax filings and execution of some of the long and short-term jobs will be done effectively by these machines. This will bring along an entirely new height of transparency, which is based on know-your-customer reporting. The intense due diligence checks performed will also be more effective and will replace the many hours that human beings take.

Even so, banks must first have a clear strategy for them to get the transformation right. There are many fields that are in the process of automation, but this might not be as easy as said. Let us look into some of the crucial functions that are likely to be transformed by AI. In fact, all of these areas have been impacted by AI, only that improvement needs to come in handy for some of them.

Risk management: data handling is quite crucial in the finance sector. AI has huge processing power, which when combined with cognitive computing assists in managing structured and unstructured statistics. Unlike humans, artificial intelligence evaluates the record of previous risks and identifies early signs of potential future problems.

Fraud prevention: battling fraud in the financial sector is getting easier as AI helps in prevention of credit card fraud, which has been the most prevalent. The algorithms can learn a customer's behavior and analyze their buying habits. In case of an uncertain pattern, the system raises an alarm. These days the banks are also employing artificial intelligence to reveal and combat money laundering.

Personalized banking: AI powers chatbots that are known to offer comprehensive solutions to clients, hence reducing workload at the call centers. There are also smart tech-powered virtual assistants that are voice-controlled and they are gaining traction quite fast. Amazon's Alexa is a perfect example and it keeps on getting smarter by day, which means there are more tremendous improvements yet to be experienced. Some of the big banks in US are already operating with applications through which their clients can interact with customer care for assistance, get reminders to settle bills, and also plan their expenses.

Credit decisions: financial institutions suffer when borrowers keep defaulting. For this reason, it is crucial for these lenders to evaluate a borrower's credit score before deciding whether or not to lend them money. Artificial intelligence provides a more precise evaluation of a borrower. Usually, it incorporates a wide array of factors, which helps a lot in coming up with a data-backed choice that is also well informed. The algorithms assist in distinguishing potential borrowers with a high risk of defaulting from the creditworthy ones who may be lacking an extensive borrowing history.

Process automation: robotic automation of processes is a sure way of boosting productivity and cutting operational costs. The industry leaders focus on such aspects like intelligent character identification for obvious reasons. Such aspects computerize a range of mundane and lengthy tasks that only blow up payrolls in the business setting. Normally, the AI-enabled program validates data and creates reports based on a variety of parameters. They also analyze documents and extract more information from necessary forms such as agreements and applications.

In essence, we cannot overemphasize the benefits of artificial intelligence in financial services; they are too many to overlook. According to a report given by Forbes, a significant percentage of senior financial administration is anticipating positive transformation from deployment of AI in financial sectors. This gives us confidence that indeed banks can effectively lead the shift to artificial intelligence.

Although only a relatively small fraction of financial organizations has embraced implementation of AI in their processes, there is a lot of hope. Certainly there is fear that the endeavor will take a lot of time and consume a lot of finances, and this is normal. Nonetheless, this is a technological wave on the move that they can't shy away from forever. If anything, holding up from embracing it now may only cost a lot more in the long run.

CHAPTER 10:

Growth of business value from Artificial Intelligence

As defined by its consequents including abilities to sense, understand and perform independently, Artificial Intelligence has attracted executives, scientists, and researchers. As a result of frequent media attention, desires have grown in capitalists to exploit the innovative business solution Artificial Intelligence assures. Accordingly, companies involved in technology have grown interested in the exploitation of AI as a venture (Bataller, C. and Harris, J., 2018). Well-contoured acquisitions such as Facebook, Google, and Apple have developed desires in Artificial Intelligence expertise precise to robotics, communication gestures and computer visualization among other fields.

Consequently, the demand for Information Systems has vividly amplified. This has been influenced by devoted desires by companies in the exploitation of these fields. However, with any new promising prospects, business executives are faced with the dilemma of differentiating hype from potential offered by Artificial Intelligence. Furthermore, as a result of skepticism and ignorance of fiscal values consequent to these technologies, Artificial Intelligence prospects face tough gusts. Nonetheless, executives should keep in accord the benefits of AI, rather than dwell in the uncertainties. In these accords, executives should appreciate the diversity and abundance of the AI field.

Artificial Intelligence: Independent, Discerning, Understanding and Acting Systems

With AI technologies, computers are independently capable to perceive, understand and act on data collected. Moreover, AI systems can learn from familiarity and adjust accordingly to the knowledge acquired.

Ability to Sense

Considering the optimization of business habituation, visual technologies i.e. facial recognition may be used to recognize; customer characteristics, and security liabilities among others. Moreover, Video analytics, an additional sensing technology may be used to cement the recognized aspects as mentioned above.

Ability to Understand

AI systems also can understand using technologies in the fields of language processing, adept systems, and interpretation engines. These technologies are widely rooted in various applications in different industries. For instance, in the medical field, the technologies facilitate the identification of diseases and offer suggestions for their treatments. These comprehensions are attained when data is keyed into the systems, which in turn follow up on the inquiries, memorize the facts, compare the facts to the knowledge possessed in concern to the inquiries, and offer breakdowns.

Ability to act

Artificial Intelligence systems act autonomously. Using technologies such as expert systems and inference engines, the AI systems can act or refer the actions to appropriate physical implementers. For instance, driverless cars act independently in accord with the environmental conditions since they have abilities to sense and comprehend numerous inputs. Moreover, industrial robots facilitate smooth operations in production lines since their actions are based on what they have independently sensed and comprehended.

The ability to learn

Rather than needing encoded instructions, AI systems can gain experience and adapt their competencies. These operations are facilitated by a technology called 'machine learning'. Nowadays, artificial intelligence systems can learn rather than rigidly accept codes that previously facilitated their operations. This coded concurrently required modification in their programs in the improvement of their undertakings.

Identification of Artificial Intelligence opportunities

In terms of automation and intensification, we may categorize and describe existing Artificial intelligence solutions. Overall efficiency and performance of AI systems may be improved through automation of routine tasks. In subjects of determining qualifications, technicalities of making repetitive policy assessments may prove difficult and monotonous for humans (Kreinczes, 2016). However, these decisions are certainly easy, swift, accurate and consistent for Automated Systems. Nonetheless, some situations can only be judged through actual human analysis. For instance, smart machines have been used in the augmentation of surgeries and fighter planes. However, it would be utterly inappropriate and impossible to automate the procedures involved in the performance of these activities. As patients would disapprove surgeries performed by 'robots', consequently automated fighter planes which operate independently and in reliance to their inhuman comprehensions, lack emotional intelligence. Therefore, the judgments taken by these Artificial Intelligence systems may not offer corporal or humane guarantees. In this accord, such fields can only require augmentation of AI systems to facilitate their mechanization.

AI Analysis Benchmarks

To maximize the outputs offered by this technological wave, businesses need to understand the criteria used to distinguish the assortment of AI solutions. Differentiations of properties comprised in the operations that need solutions, affect the preferences of AI opportunities. Whether automatable or augmentable; AI solutions need to weigh the density of the task in the operation and also, the density of the facts and evidences involved.

Consequently, the nature of the work may be predictable and direct, while the spectrum of the activities involved may be unknown and unpredictable. This consequence oversees the importance of human inclusion or augmentation of the systems involved in the actualization of the operations if the systems support. However, sometimes data may be constant, reasonably structured and in small volumes.Subsequently, the frameworks gathered in the accords of work and data complexities, four prime forms of activity prototypes are evolved. These models include; innovation, expert, efficiency, and effectiveness.

Innovation model

Artificial Intelligence technologies facilitate the identification of alternatives and optimization recommendations, while humans make resolutions and act on them. In this model, AI systems bring out creativity and human inspiration.

Expert models

The work involved in this model entails human judgment and is conditional to human familiarity and capability in the undertakings executed. Consequent to the fact that decision making and exploitation is chiefly a human duty, technology is only involved in the augmentation analysis procedures.

Efficiency model

To optimize business performance consistency, the efficiency model characterizes routine undertakings defined by instructions, processes, and rules. In the solutions involved in this model, technology discerns, understands and acts, while humans observe the precision of the solutions developed and regulate the evolution of the business environments.

Effectiveness Model

This model facilitates the improvisation of general employee and business abilities to yield the desired effects. The efficiency model informs workers on all their businesses entails. The success of this model is earned through communication and organization successively affected by aspects such as administration and sales. The actualization of these solutions may require technological interventions, where the AI systems act as assistants or agents.

Challenges of Artificial Intelligence

Human discomfort about AI Abilities

Some humans are threatened by the capabilities of AI creations. Due to their lack of emotional intelligence, their decisions may seem threatening to humanity. Therefore, businesses should be selective of their AI solutions to ensure the experiences of their employees and customers are favorable and unsuspicious

Purveyor Hype

As a result of previous unfruitful prospects of Artificial Intelligence systems, several businesses are justified to be wary of their involvement in AI systems. Developers should pitch their technologies with precision, expressing both the benefits and liabilities of the prospect. Moreover, the companies should familiarize themselves in undertakings that enhance the understanding of the solutions, before and after the acquisition.

Differences in Expansion Methodologies

Previous systems mainly based their developments on planning, analysis, designs, testing, and disposition. However, AI environments dwell on identification and purification of the research to suit the desired output. To elevate approaches in AI environments, different methodologies, abilities, and approaches need to be integrated into the environments. Moreover, intellectual systems need training in specific areas to ensure their effectiveness(Askary et al., 2018). Therefore, to attain suitable development models, AI requires different expansion methodologies such as; teaching, instructing and managing the solutions it offers.

Humans vs. technologies

As pitched by several movies, new technological ventures will create abominations that will overturn our superiority as humans, which will, in turn, lead to our enslavement by the artificially intelligent machines. However, and as previously highlighted, these solutions can exempt humans from routine and monotonous tasks, facilitating the concentration of the human workforce to more important developments. Moreover, such solutions facilitate the augmentation of other cases that require the preservation of emotional intelligence in their performance.

Societal insinuations

As new technologies evolve, various social aspects such as employment are disrupted. Therefore, different themes are experienced in the takes societies have on AI solutions. As coined by the Pew Foundation (Russell et al., 2016), societies will adjust to new ventures that only require human uniqueness for their undertakings. Moreover, the collaboration between humans and machines may certainly benefit both parties since they can learn from each other. As Google's Larry Page affirms, the internet is an example of a mutual relationship that benefits human lives and integrates new knowledge to the internet.

Artificial Intelligence develops businesses across all industries. However, the most important aspect to remember is to avoid being absorbed to the notion that any technology is the answer to everything (Banerjee et al., 2018). It is important to weigh the vitality of the works to consider the best rationale for the incorporation of technologies into complete AI solutions. Continuous discussions on the effects of these technologies on existing human activities will hasten the incorporation of AI solutions to businesses. A joint intervention of businesses, educators and executives are needed to extensively evaluate the pros and cons in AI solutions, and act accordingly. The evaluators and creators Artificial Intelligence solutions will be liable to any social effects, good or bad; will not be blamed on the emotionally ignorant technologies. However, the benefits of Artificial Intelligence are more significant to their complications.

Chapter 11:

Chatbots and autoresponders

AI powered by digital solutions to improve customer service is a new revolution, which every industry worth its salt is engaged in an attempt to transform conventional customer service in every aspect including product knowledge, brand awareness, customer acquisition, loyalty programs and after sales service.

What Are Chatbots?

A chatbot is a kind of Artificial Intelligence (AI), which uses an automated robot system that creates a virtual conversation through text chats, voice commands or both, so much like human conversations, in order to relay automated replies or carry out specific tasks for users based on a well laid down set of rules or parameters.

Secondly, a chatbot can be used on messaging platforms, websites or applications where it can perform many functions.

Chatbot, also referred to as Talkbot or Bot in short can also be defined as an Artificial Intelligence computer program or simplified software application, designed to imitate or simulate either spoken or written human conversations or interaction with individuals, mostly through the internet in an automated manner.

It is said that the demand and usage for these messaging applications has outwitted even the most reputed networks of the social media, dating back to the year 2015. This indicates that as popularity for messaging Apps increases, more and more consumers will desire to use chats to for interaction with services and brands. By the year 2017, publicity for automated conversations increased greatly, altering brand conversations with end consumers. A good example is Facebook Messenger, which in the year 2016 broke the record of active users by a whopping 900 million.

Business customer service and sales can greatly improve through use of the now familiar messaging chatbots interface leading to heightened customer interest and consequently increased earnings and profits

How does a chatbot Work?

You might be curious, like many are, on how chatbots work. This is very important since the current trend indicates that in the near future this piece of automation system will play a key role especially so in the business arena. As a business person, you may want to try this out in planning your business as you align it with your precise business needs.

Two types of chatbot software Applications in the modern times are Artificial Intelligence Chatbots which processes natural language. This helps the chatbot application recognize and comprehend text messages or human speech, as well as discerning intent. An example is when

one sends a message requesting for something, the system is "smart" enough to recognize and understand the meaning. More interesting, it even responds by asking follow-up questions such as "do you want tea or coffee?"

On the other hand, Rule-based chatbots is whereby specific commands are used in order to receive a response. This chatbot category denotes a major opportunity for enhanced customer sales service and profits for example use of some text messaging App used by some retail businesses to send offers or coupon codes advertised with a sign on a coffee shop reading "Text COFFEE to 21534 for a 5% offer."

Another example is when your phone texts "COFFEE" through the messaging system, the App recognizes the word as a command, then follows the developer rule and instantly sends back to you the coupon code. If however the word "COFFEE" is misspelt it might not pick.

Chabots can be said to be an Application without the User Interface. To understand better let us compare an e-commerce website functions to that of a Chabot. The former is a software APP for selling products. The layers or aspects that make it work are:-

An Application layer: A set of instructions which function-ability to the APP.

Application programming interfaces (APIs): Link the App to services so as to get for example process payments or shipping quotes.

User interface: Enables the user to inform your application what his interest is or what he wants.

Database: Stores customer data, product information, bulk of the content and transactions.

The user interface for the e-commerce website therefore plays a key role in the function-ability of the App. It has for example a cutting-edge search tool for locating products by shoppers at the same time displaying incredible product pictures. It has user buttons linked to a shopping cart where one can add products. It also has templates for entering address and payment info.

On the other hand the latter (chatbot) powered by natural language processing and Artificial Intelligence has three of the e-commerce aspects, including the application layer the database, and access to Application programming interfaces

APIs). It however does not have its own User interface but depends on a messaging platform. It can for example use Slack, Facebook Messenger, WhatsApp, or other comparable App for customer or shopper interaction. This means that the developers therefore must write links to a number of other platforms as mentioned above so as to potentially achieve a reach similar to an e-commerce website.

What are autoresponders?

An Autoresponder refers to a service that automatically enables one to send emails to different groups or a single group of people.

It is also an effective yet simple marketing tool which is used to send a succession of scheduled emails based on set specific parameters. This is manly used for follow-up purposes on ones shoppers or customers within ones store or business.

For promotion of specific items and also for new product demand tracking purposes an email can be sent to all customers or shoppers, for example as a follow-up. This can be in form of a voucher or special coupon or simply informing them of other similar products or related accessories that would perfectly rhyme with what they have bought.

How do Autoresponders work?

Autoresponders are a powerful marketing tool that is very effective and essential for internet marketers and business owners. They enable one to contact innumerable potential customers. If you consider the number of emails you are going to make just to make a single sale, having an autoresponder is a great asset for your business. Autoresponders automate up to 50% of one's marketing campaign, without which a business could lose probably a bulk of sales per annum.

Email marketing tools and Autoresponders are very important business components as follows: -

Automatic email sending

Automatic email response

Used in following up with customers.

Sending product info to clients eg. price lists and offers, newsletters

To build loyalty

To keep people in constant connection with your site.

To make sales.

Creation of email courses

Once emails are written, they are sent automatically, then subscribers sign up to receive.

Some commercial autoresponders, in addition to sending out standardized messages can send limitless follow-up messages to members, instantly without time interval delays.

Some hosting companies can provide a free autoresponder, however, purchasing an independent service creates a wider scope, to carry out more tasks like, like personalizing emails, where one can include name, phone number, address, business name, which enables attachment of a business card.

Autoresponders allows creation of short email courses filled with relevant information, running for 10-15 days which can be offered to ones' visitors for free which will bring in targeted customers depending on the field of expertise selected. Article briefs are auto sent to the visitors every day. This will also create reputation for the company.

Examples of Autoresponder tools which you can research on are:

1. Office Auto Pilot
2. Campaign Monitor
3. Aweber (used for creating and sending email newsletters for marketing purposes)
4. Infusion Soft
5. Auto Response Plus (ARP Reach)
6. Mail Chimp
7. Get Response
8. iContact
9. 1ShoppingCart
10. Constant Contact

Persuasion through AI, Automation and data science

Persuasion can be defined as persuading, a term of inducement or influence. A person can be persuaded by attempting to influence their behavior, attitudes, motivations, beliefs and intentions. In the business arena on the other hand, the process of persuasion is aimed at influencing or changing the behavior or attitude of an individual or group towards an idea, object, event, or towards other persons, using spoken, visual or written words to convey information, reasoning, feelings or a combination of the three.

Persuasion is a tool that is oftentimes used in quest of personal gain such as campaigning for electoral position, in sales promotion, or advocacy. Persuasion can also be described as using ones' position or resources to change people's attitudes or behaviors.

Heuristic persuasion is whereby beliefs or attitudes are leveraged by emotion or habit.

Systematic persuasion is whereby attitudes or beliefs are leveraged by reason and logic.

Automation is a technological procedure or process performed with nominal human assistance. Also referred to as automatic control uses a variety of control systems in equipment operation. Equipment like machinery, factory processes, ovens for heat treating, telephone networks boilers, stabilization and steering of ships and aircraft, vehicles with reduced or minimal human intervention and other applications.

Automation includes applications ranging from large industrial control systems to household boiler controlled by a thermostat, from high-level multi-variable algorithms to simple on-off controls.

Automation has been applied mostly in combination in mechanical, pneumatic, hydraulic, electrical, electronics and computers. Complex systems, like airplanes, ships and modern actories, and typically apply these combined techniques. Automation benefits include savings on labor costs, savings on material costs, savings on electricity costs, and improvements to accuracy, quality, and precision.

Scientists help in retrieval of valuable information from a sea of data to be examined and the discoveries used to streamline the companies. The role of data scientists is to analyze data, asking information-driven questions, applying mathematics and statistics in order to discover important outcomes.

In our current digital world where there is a remarkable increase in automation, there is an urgent need for us to initiate conversations and dialogue, especially so in recent times where we are confronted with the amazing field of Artificial intelligence. What with smart voice chatbots, autoresponders, Netflix, self-driving cars and the field of robotics, just to mention but a few. For us to embrace Artificial Intelligence (AI) for example, we have to be made to understand not only how the different components work, but how it affects society. Oftentimes, persuasions regarding automation are mostly centred on the job market, its negative or positive impact, more so as AI systems become increasingly commonplace.

Two ways in which the current trends in automation will influence and affect our future is that for one, it will in no doubt lead to a more efficient and superior future as witnessed by industrial revolution- jobs will be lost yes but due to rapid expansion, new openings will occur for better more advanced and efficient jobs.

The other way is to recognize that the present times are extraordinary with robots becoming increasingly fit and smart. It is possible that the number of job opportunities and enterprises they will consume or crush will outweigh by far the number of job openings they will create.

Whichever of the two possibilities above turn out to a reality, what we know for sure is that automation or AI for that matter is drastically influencing the economic landscape in these extraordinary times. We are in realization that a big percentage or portion of our lives and jobs are increasingly being automated.

This concern is not new for during the nineteenth century, textile workers in England are in record to having destroyed weaving looms, these workers were then accused of being opposed to technological advancement.

The talk is rife of the possibility of intelligent machines in future turning against humans, or their makers for that matter. Potential for disruption of the labour markets by intelligent machines has been examined through academic studies and various books published most recently. Furthermore, detailed studies have been carried out on a possible serious threat posed by automation to human existence altogether, with special reference to AI.

The AI technology could not have been predicted better than through a classic science fiction Space Odyssey film of 1968, directed and produced by Stanley Kubrick. The film, following a space voyage to Jupiter discovers beneath the Lunar surface a mysterious featureless artifact or alien monolith. Mankind then sets off on an expedition to discover his origin using HAL 9000, an intelligent sentient (perceptive and intelligent) supercomputer.

Automation is nothing new of course for robots have been in use for decades in industrial assembly lines. The current automation wave draws its benefits from affordable computing power with emerging new areas in software application such as image and language processing. This new generation automation machines and AI systems may adversely affect white collar jobs which was not previously so. Apart from potential consequences to employment, there are other widespread implications in other areas such as education, privacy, cybersecurity, healthcare, environmental management and energy.

Cheap bandwidth for example is influencing learning methodologies which can be seen through online platforms which is greatly influencing the how and what is learnt. A good example is MOOCs (Massive Open Online Courses). As automation increasingly takes a big chunk of routine tasks, this kind of learning which stimulates creative and conceptual capacities becomes more relevant and popular, consequently leading to a shift in the education system from conventional reading and mathematics to more focus on intellectual and personal skills, working hand in hand with intelligent machines akin to AI.

Universal individual and commercial data collection and storage from social media platforms on the other hand raises privacy concerns. Yet another area of risk is Cyber-security. Developed economies are feeling the impact of automatization more than developing countries which are following closely behind especially so through investors.

Customer service: reduce workloads and handling retention, segmentation and scoring

Two benefits of application of AI in customer service is workloads reduction and customer retention

There is increasing demand globally for excellent customer service due to exposure to more advanced products and services. Customers' expectations are rising demand for seamless service is increasing day by day, especially so with more and more automated services that are technology aided, social platforms, messaging services and applications.

Customer retention is key as it boosts revenue, thus businesses target is to build loyalty and in so doing retain customers. It goes without saying that good customer service leads to customer satisfaction which translates them into brand ambassadors. They carry out free brand promotion to family and friends, consequently boosting sales which increases revenue.

It is proven beyond any reasonable doubt that poor service sends away customers and on the flipside they return if they receive good service consistently. Customers are prone to adept and efficient systems and cannot prefer an inefficient, outdated system when there are alternative advanced channels which are faster and time saving akin to automated systems such as AI can offer.

Artificial Intelligence (AI) therefore combines intelligent services with automation in order to achieve accuracy, scalability and efficiency which manpower alone cannot achieve.

AI can be tuned, trained and customised to fit business processes in the current times. AI-enabled customer service from contact centres improve or enhance human agent capabilities supported by AI chatbots provide customers with a wholesome service experience. AI chatbots can handle common customer queries and even suggest solutions while human agents can handle more complex ones.

Predicting Consumer Behaviour and Referral programs

With emergence of Artificial Intelligence (AI) could help simplify understanding of consumer needs and wants ideally even before they even do themselves. Deep learning, being a sub-set of AI can potentially transform marketing in supporting prediction of consumer behavior by businesses. This machine method of learning uses deep or layered neural networks, comparable to biological brains able to not only acquire or learn skills but also in solving complex problems at a higher rate than humans. It helps robots or computers to perform "human" tasks like recognizing voices, translating languages and perceiving objects. With a set of inputs, deep learning helps train AI in predicting outputs. However deep learning requires large data and computational power, though much less data pre-processing by humans is required as compared to machine learning techniques. With access to key elements however, a deep learning system can learn human behavior prediction pretty accurately.

Referral programs are becoming a widespread way of acquiring customers. However there isn't much to prove that these nature of customers carry more weight than regular customers. The question still lingers whether the referred customers are more loyal and more profitable and if so, to what extent? Researchers have attempted to determine this.

A leading German bank was tracked approximately 10,000 customers for a period of about three years. The findings were that referred customers:

1). Show a higher contribution margin rate which however eroded slowly by slowly

2). Indicate a higher rate of retention that the regular customers which persists over time

3). comparatively in the short and long run they are more valuable than the counterpart customer of similar time of acquisition and demographics.

A referred customer recorded higher average value by 16% than the non-referred customer

Up-sell and Cross-Sell with AI-Powered Product;

Upselling refers to a product promotion using a complementary option for example after one buys a computer the customer is given an offer of alternative memory options or processors ("we have a higher caliber computer than your selection, would you consider adding some little cash you have it at a discount...."), while on the other hand in Cross-selling the customer is encouraged to purchase a different product with their original purchase, say accessories ("can you also take this flash disc at a discounted rate which is very crucial...?). In cross-selling, the complementary products do not have to be those that work together with the purchased product one can suggest related products not necessarily those that work together. Sales services too such as priority service or warranty is also a cross-selling strategy

Use Predictive Analytics to Provide New -Product Recommendations and Boost Revenue, data-driven approach

Predictive analytics is a combination of predictive modelling, advanced analytics, real-time scoring, data mining and machine learning which helps identification of data patterns by companies. Historical data is used to predict future trends in business for example. This is achieved by feeding the historical data into a model which analyses the data and identifies patterns or trends. The model studies the historical data, which is then used to predict or forecast future outcomes by applying it to current data.

Predictive analytics is currently used widely in many ways such as: -

Product recommendation system which predicts customers' likes

Customer Lifetime Value (CLV) measures predict the estimated amounts a customer will purchase form a company

Others are; Predictive maintenance; Optimizing marketing campaigns, credit scores, fraud detection and sales forecasts

According to an Accenture survey conducted recently, it shows that since 2012, most company executives have placed predictive analytics high on their executive agenda. This is due to increasing awareness on predictive analytics indicative of potentially profitable future outcomes.

How do predictive analytics work?

With large amounts of data, the way relevant insights are derived from information has greatly changed. The approaches previously used referred to as Business Intelligence (BI) which used typical structured data sorting technics have been switched to techniques which use raw information. While the traditional BI uses deductive approach where the assumption is that there is some relationship or understanding between existing patterns with relationships, the Inductive approach on the other hand is more concerned with data thus doesn't make any presumptions about patterns, relationships and discovery.

Predictive analytics thus use the latter- inductive reasoning applying it to big data through machine learning, Artificial ntelligence and neural networks, to identify patterns and nterrelationships.

Predictive analytics can be used in: -

Business problem solving

Decision-making

Optimizing processes

Cutting down operational costs

Improving customer experience

Market opportunities identification

Risk reduction and mitigating through problem prediction.

In a company's sales department for example, staff ordinarily rely on their expertise, knowledge, professional experiences and even intuition or "gut feeling" in decision making. For example, they know their regular customers off head, customer's interest or preferences, which products move and so on.

Some or most of this knowledge is based on guess work through intuitive choices or even personal preferences.

If such a team therefore is enhanced with AI, actual data is used to help them come up with more accurate and efficient predictions which in some cases confirms their previous intuitive choices to be true. In many cases they will watch with awe on huge quantities of data unlocked by automation or AI for that matter which would have ideally been omitted.

Another advantage is that in the event where an expert leaves, which is usually the case, the AI model stays together with the unlocked knowledge so no need to panic.

CHAPTER 12:
THE RIGHT ARTIFICIAL INTELLIGENCE APPLICATIONS FOR YOUR BUSINESS

It's very important to understand how Artificial Intelligence can be of positive impact in your business. There are many AI applications that can be used in businesses to make running of the business much easier. These are several Ai applications that when suitably applied can play a key role in reshaping of your business.

Automated systems that are installed with AI can greatly help you to make better decisions in managing business resources.

There is a wide array of Artificial Intelligence applications that can be employed in business for beneficial purposes. Some of the Artificial intelligence applications are used in different industries including; healthcare, customer support, finance, security, drones, smart cars, creative arts, among others.

In the present day, most of the customer questions or complaints are responded to by human beings; this is done via telephone calls, emails, social media dialogues and online chats. However, AI applications have been absorbed in business systems which has facilitate the automation of these communication. Computers may be installed with Artificial Intelligence applications to enable them to respond accurately to customers. Further, combining AI and ML enables the AI machines and computers to work more efficiently. Choosing whether or not to use AI in your business will be determined by what you hope to achieve, in the long run.

How are artificial intelligence applications applied in accounting in business?

Business accounting refers to the systematic recording, analyzation, interpretation and presentation of financial information. In small capacity businesses, accounting is usually handled by one person, or by various teams in larger organizations.

The accounting process is of great importance in any business in that it helps the business owner to keep track of the business' operations. Accounting process helps in analysing finances which enables the owner to make wiser decisions, especially in handling finances. Incorporating Artificial Intelligence applications in businesses makes the accounting work easier, which enables the business owners to honour their compliance obligations.

Artificial Intelligence applications are set to transform the accounting and finance industries; there is the elimination of tedious tasks which translates to saving time. Employees can therefore engage in other tasks that have more productive impact on the business.

HANA is an Artificial Intelligence application that helps businesses to convert their database into intelligence tools that are more useful and practical. The application works by ingesting and replicating sales transaction data drawn from relational applications and databases.

In drawing deeper insights and streamlining finance processes, firms need to explore current AI applications. These applications assist accounting managers to stay ahead of business transactions in the midst of time-consuming and tedious processes. This is made possible by making use of the Machine Learning app instead of spreadsheets and PDFs. Receipt images are extracted and automatically classified by the Machine Learning app; the extracted data is classified bases on the spend category. These reports offer smart insights for businesses which help in better financial planning.

The ML app is a branch of AI and it yields deeper insights in that data processing is done over time; this provides businesses with comprehensive overviews on the spending patterns which can be used in making better finance decisions for companies. The basic AI apps are undoubtedly key players in the business accounting field.

Are Artificial Intelligence Applications used in founding for startups and established companies?

Startups have been dominated by various software giants including SAP, Oracles, IBM and Salesforce. They provide various applications that are applied in customer relations management, resource planning and human resource management. Several startups are offering the next group of innovative services via Artificial Intelligence apps,

Companies that are powered by Artificial Intelligence, many of which are AI startups are making use of these in their recruitment processes. For example, Salesforce has made investments in DigitalGenius, which is a management solution for customers and in Unabel which offers translation services for businesses.

Artificial Intelligence startups are now generating solutions for various industries. These can be seen across several business sectors i.e. legal, automotive, agriculture, healthcare and finance. It's very clear

The Artificial Intelligence startups are becoming very popular because of their capability to provide established companies with prized point solutions; this is mainly due to the fact that they can access domain knowledge and large data volumes.

A majority of AI startups are developing some type of Machine Learning models that have the ability to predict or classify end results based on certain input data. In majority of AI startups, there effective running is highly dependent on data volumes; the larger the data volumes, the better their performance.

Artificial Intelligence in Creating the right team to lead success in business

The rapid increase of Artificial intelligence in business has been brought about by the need to answer various questions in the recruitment process i.e. the exact skills you are looking for as you are recruiting, the organizational structures that will work best for your business, the organization module that will result in success while using Artificial Intelligence, etc.

Many businesses have realized the importance of implementing Artificial Intelligence teams that play important roles in improving productivity and service delivery to customers.

Making use of AI teams that are well-rounded enables businesses to reap maximum AI benefits. In the process of selecting members for the Artificial Intelligence team, it's important to consider reorganizing your recruitment processes.

You will require talent that can grasp computer learning processes. A centralized team is basically what will be the driving force behind the Artificial Intelligence applications. Always remember that the main aim of Artificial Intelligence is to basically reduce work in the shortest time possible. Having he right team in place will facilitate better performance of the business which is bound to reflective positively in the long run.

The effect that having the right team in place in any given organisation can never be underestimated. This is why employers need to employ the useful AI apps to enable them to recruit people with the right skill which ensures that tasks are performed within shorter time frames which can greatly help in meeting targets and improving productivity.

The Artificial Intelligence team is tasked with executing the whole AI project successfully; this means that the team must have engineering talent for them to carry out this task successfully. Depending on the project that needs to be implemented, data engineers, data scientists, machine-learning engineers and product managers could be among the AI team members.

Baidu and Google are excellent technology teams; they have proven that they have the capability to function cross-functionally in creating strong AI value via advertising, product recommendations, speech recognition, etc.

How are Artificial Intelligence applications used in the collection, preparation and recycling of data?

Data is generated in huge volumes in the business world. These can be either free text, categorical or numeric data. Data collection can be defined as the procedure of gathering information and measuring it from several different sources. If we are to utilize the collected data in problem solving in our businesses, there must be sensible collection and storage of the data. These will facilitate the development of practical Machine Learning and Artificial Intelligence solutions.

The procedure of data collection and preparation is very vital in business since it enables you to have records of past events; the data analysis then kicks in when looking for recurring patterns. Predictive models are then built from these particular patterns which can be used in predicting possible changes in future.

Good practices need to be observed when collecting data to facilitate the development of high-performing models; this is highly dependent on the data that is collected. The collected data must be free from error and contain pertinent information to facilitate the performance of the set task. The collected data is then re-used for laying strategies for solving similar problems or tasks in future.

Artificial Intelligence analytics solutions for companies
Artificial Intelligence analytics solutions possess the means through which value is derived from the rich information generated from the collected data. Progressive analytics platforms enable businesses to create insights on business processes as well as predicting outcomes; this is based purely on information that is collected from data in the past. Artificial Intelligence on the other hand incorporates Machine Learning capabilities by acting as the force multiplier for the generated insights.

Artificial Intelligence applications and analytics solutions aim at making businesses more productive, effective and efficient by providing valuable enterprise insights in an accessible manner. These insights are linked to the most important goals of the enterprise.

The best analytics software and AI provides leverage for ML algorithms in data platforms which transforms big content and big data into data visualizations; these help the business to increase operational efficiencies, maximize revenue and increase automation.

Open Text Artificial Intelligence and Analytics provide a detailed set of analytics tools and AI which help businesses to sort out data challenges at a comfortable pace. This warrants a clear path towards Artificial Intelligence technologies whose main focus is on improving decision making, business optimization and automation. This can only be achievable with the appropriate tools for generation of insights. The main achievement is the identification of trends, patterns and relationships through analysis, processing and recycling of data. Businesses become better equipped to handle similar situations in future which leads to improved productivity for the enterprise.

Some of the positive impacts of Artificial Intelligence and analytics platforms include:

- Increased operational efficiency
- Cost reduction
- Increased productivity
- Accelerating the rate at which AI value is delivered in a cost-effective, time-saving manner
- Improving decision making and visibility
- Maintaining operational improvements continuously

Business strategy in the Artificial Intelligence economy explained

Breakthroughs in the business economy related to the artificial intelligence program has led to the redefining of business operations across the world. Business strategy and Artificial Intelligence initiative has encouraged the growth in the use of AI in the enterprise landscape. Specifically, this exploration looks at the manner in which AI affects the execution and development of business strategy.

This particular initiative is aimed at reporting and researching on the manner in which Artificial Intelligence has resulted in a drastic workforce change, privacy, data management, the generation of new business challenges, and cross-entity collaboration. It investigates new threats and risks in security, job loss and dependency as well as seeking to assist the business managers to understand the huge opportunities from combining machine and human intelligence.

An example of the manner in which business operations have been refined is the manner in which data is collected, processed and used in the optimization of customer interactions. The alteration is playing a key role in the redefinition of structural dynamics of the management concepts, economic theory, and business strategy. Leading companies in the technology field globally have cast their nets wider in their research with regard to Artificial Intelligence.

AI has the ability to help in transforming businesses just in the same manner that the internet has transformed the way businesses is done. From smarter services and products to automated and optimised business processes and better enterprise decisions, Artificial Intelligence is well able transform everything in the business world. Businesses that aren't capitalizing on the Artificial Intelligence transformative power risk being excluded in the growth process.

In conclusion, there are several Artificial Intelligence applications that are in use in businesses. The main aim of implementing is the businesses is for the possible benefits associated with me. It's critical to do a proper analysis of your business and the Artificial Intelligence applications to determine which apps will work for your business and those that will not.

Cost implication is the major concern; many businesses when at the implementation stage of Artificial Intelligence lack adequate personnel. This mean that they have to outsource for manpower to ensure that the system delivers the intended results. Another setback is the cost implication in regards with the maintenance and implementation of the AI system. Sourcing for AI applications that are pocket friendly is a great idea in cost reduction. Take time and find out the applications that are being used in businesses that are similar to your business.

CHAPTER 13:
ARTIFICIAL INTELLIGENCE FOR
MARKETING AND ADVERTISING

Artificial Intelligence in Marketing and advertising: a revolution for your business

Artificial intelligence marketing is a technique of grasping client data and AI models like machine learning to predict your client's subsequentaction and to improve the client 's experience.

Key components of AI Marketing

AI based marketing is a very powerful tool today that we cannot ignore. It comprises a couple of aspects as listed below.

Big Data

It refers to a marketer's ability to combine and fragment huge groups of information with least labor-intensive effort.

Machine Learning

They are crucial when analyzing large volumes of information. This is by pointing out patterns or repetitive occurrences. Also they help in accurately foretelling ordinary acumen, feedback, and opinions. This in turn helps marketers comprehend the origin and probability of certain undertakings recurring.

Powerful Solution

This means that they identify intuitive ideas and themes within large data groups, extremely fast. Also they deduce feelings and communication like a person, which enables them to comprehend information from social media, and email responses.

Digital optimization and personalization

Digital optimization is the action of using arithmetical technological expertise to advance current working routines and corporate designs.

The progression of **hugerecords** together with progressive logical results have made it promising for venders to form a vibrant image of their goal audiences than in the past; and in this hotbed of advancement lies artificial intelligence (AI) marketing.

Equipped with large data intuitions, arithmetical salespersons can significantly improve their crusades' results and the return on investment. This can be realized with fundamentally no further work on the Salesperson.

What marketers do better than machines

Once it comes to numerical breakdown and quantity chomping, technology is the best option if you're operating *with* it, but then again your worst rival if you're contending against it. Technology is faster and more precise than people, and its inferences cannot be influenced by individual or emotional preference.

The theoretical inference is that we ought to turn most of our choices, likelihoods, diagnoses, and verdicts, both the inconsequential and the consequential, over to the procedures. There's is no argument at all about whether doing so will provide us with superior outcomes.

Why people will keep on being essential to marketing

We are not existing in those revolutionary sci-fi cinemas so far! Apparatuses are compliant, we authorize them whatever we want them to do, and don't have a brain of their own, so they definitely don't have the expressive intellect to attach with a human.

Machines can't distinguish what's "hot" and anything's that is old. Movie marketing is a picture-perfect illustration of this; a device remains incapable to perfectly guess which movie character a precise individual will hunger published on their outfit, based on a trouser they bought in the previous summer. Human specialists use a combination of their personal emotional intellect and analytics to guess this. Some people's likes and favorites are puzzlingly qualitative and unplanned. They are frequently not demarcated by relative groupings, but reasonably by random and unscientific aspects, the summation of a lifespan of irreproducible involvements and the memories that linger. This marks such a guess unachievable by a machine.

Machines do not have a soul

How does one recognize when someone has taken their time to do something? You don't even know how, but one can make out even the well-crafted computerized email. How do you tell? It's probably be the lyrics used, the sentence structure, or even the layout, but one way or another you just discern it's a machine. In some way, we comprehend the emotional limitations of technology, and this makes us important to communication with our audience

In what way is marketing technology intended to support humans?

Technology isn't intended to substitute human, it's intended to help you as a marketer. When you recognize how to leverage technology, you will join the greatest, fruitful and profitable products.

As a matter of fact, advertising technology is undertaking the hardest part of your marketing progressions, like segmenting your clients and diagnoses and directing them with content that is contextually applicable to them at a certain time. This gives you the chance to turn your devotion back to the skills of influence, and to get involved with your clients on an expressive level that apparatus basically can't.

Here are several vital takeaways one ought to know about where this expertise can go and in what way it can benefit you.

Machine knowledge will at no time substitute humans as the leading spring of inventiveness; it will simply assist make that content manufacture further effective.

The total amount of content and information to be created in he subsequent number of years will exceed the human brain capability to create and compute.

Machine knowledge will offer products with the supremacy to put their Huge Information to use and control their analysis rather than allowing the data slip up through their fingers.

Machine knowledge will aid in creating a more modified involvement for the operator, refining his or her impression of he product.

It will assist brands create their marketing hard work more effective by enhanced targeting clients.

In spite of that, there is still a crucial requirement for human inventiveness and awareness, meaning there isn't an abrupt danger of digital dealers becoming outdated. Yet, when approaching the overview of fresh technologies, it's essential not to oversee administrative fundamentals and the human level in edict to ensure operational implementation and operative satisfaction. Those firms that have moved fast to revolutionize the whole client expedition by leveraging modern t technological advances, whereas also maintaining the human level in mind, will be the ones to win the contest for digital mystery.

The correlation amid humans and machines has transformed drastically in latest years, and will keep changing swiftly in coming years. It's at a critical moment at which technology has been extensively accepted as the enabler of commercial progression and fast-paced invention, letting us to gage processes and manage high capacities of information.

Marketing platforms and tools for your company.

Before choosing AI based marketing platform there are a couple of factors that you have to look out for. They are as follows

How user friendly is it?

You need a marketing platform that is easy to use and navigate. This is because as a marketer you need to spend as minimal time as possible maneuvering a said platform. It saves time and is easy to learn and master. Also consider the users and how well they interact on the platform in regards to timeliness delivery of reaction, response and feedback.

s it compatible with other marketing platforms that ou have?

efore you purchase any AI marketing product you need to onsider other products that you already have. How compatible s the new product with the pre-existing ones? This ranges from he content management systems (CMS) to any email narketing software.

s a company, is there enough data to research from?

'he efficiency of any AI platform is contingent on the volume of ata that is fed to it. The more data there is to research from the nore accurately will it address your problem.

Iow well does it address your particular problem?

ven before sourcing for an AI product are u certain that you ave narrowed down your specific problems and if so how ccurately does this product cover or solve this problem

Vho are the developers of the product?

n these times of inventions and new discoveries there are a ton f developers out there. However, when it comes to settling for n AI product,you need to research on the developers and be ure that they are people who are conversant with the market nd its' evolving trends.

Iaving the above in my mind, below is a couple of marketing latforms and tools that may be suitable for your company. We will divide them into five entities that are based on the purpose and area of use of each.

Artificial intelligence product for content generation

Content generation process is key to marketing. It is not an easy task and marketers can attest to it but not to worry since AI products are now available. They come in handy by scrutinizing existing content and find data based methods of making it better. Some of the products under this category are;

Acrolinx

Acrolinx assist you in making certain that the content generated matches with your strategy. You feed the tool with any instruction that you have generated, and it gives real-time feedback. This feedback is to confirm whether your content matches the instruction. Eventually, the product researches from the content it scrutinizes and is able to give better ways for how to better the content you create.

Brightedge

Brightedge is a Search Engine Optimization that has an AI fueled attribute called Insights. With this tool you get better results. This is because of its ability to go through tons of web pages providing precise action items to clients they to improve their SEO, including suggestions on methods to update content creation.

Artificial intelligence product for content strategy

Content strategy development is key to good content marketing efforts. This is done with a combination of human intuitions and data that is effective. By using AI platforms marketers can now analyze data to generate more precise facades, realistic goals, and more formidable content plans.

MarketMuse

It is based on learning and natural language processing. This is by scrutinizing a corporation's data comparing it to multiple websites based on the same content. From the analysis it gives suggestions for content improvement as it is generated.

Concured

The product uses data and machine learning to recommend content topics founded on audiences reacts to responds to, finds loop holes in already available content in the market and give guidelines what content to promote.

AI Products for Personalization

Individuals progressively anticipate for a tailored marketing experience . But there are some AI products predominantly fixated on using machine learning to aid salespersons convey experiences to buyers tailored to their interests.

Drift

It offers AI-driven chatbots—an altered way to generate a tailored experience for guests to your website. Using natural language processingmachine learning chatbot can decide the most appropriate responses to the enquiries guests have in actual time.

Uberflip

It uses AI to decide which of the content fragments that will be the most suitable to each guest centered on data about their actions and interests. The tool programs the procedure of generating an appropriate experience for respective individuals that frequent your website.

AI Products for Email Marketing

This is one of the most vital virtual marketing strategy. This is due to its remarkable ROI (return on investment).

Optimail

It is an AI product that helps you describe your key email marketing goals.

Phrasee

It is an AI product specifically dedicated on writing better subject lines.

AI Products for Promotion

On any website you need as much traffic as possible. With AI tools can now help you get improved outcomes for your promotion campaigns.

Cortex

It's an AI product that evaluates loads of data topics from different companies studying what the public react to. From this information, organizations can now know when and what to share on social media to achieve more impact.

Albert

It's a self-learning marketing platform that programs the conception of promotion campaigns centered on previous data.

Acquisio

It's an AI tool for handling PPC campaigns. By using AI, the campaign creation process is simplified maximizing campaigns for better ROI.

PERSONALISATION OF CUSTOMER EXPERIENCES

The capability to generate appealing connections through digital networks with clients has a major effect on progress, returns and product activism. It is accurate to note that feelings can be applied to monitor corporation digital scheme for building digital network associations with clients.

AI is aiding convey improved consumer involvements by creating improved merchandise recommendations, aiding quicker sales and enabling more appropriate personal shopping.

ANITICIPATING CLIENT PROSPECTS.

What better way of conveying improved involvements than by forestalling clients' hopes and provide exactly their desires in a timely manner.

Various businesses use merchandise commendation engine to find most consumed client favorites and merchandise bring them to a brick and mortar setting.

CUSTOM-MADE COMMENDATIONS.

Clients who are tech-conscious customer require personalization and vendors are rising to the task AI, machine learning and large data.

AI can suggest products based on recurrently purchased things or similar products.

CONVERSATIONAL ADVERTISING

Conversational marketing is basically a feedback focused on marketing tactic embraced by enterprises to motivate client base and client reliability.

For a business to mature and continue being important, they require to establish a smooth link with the customer, influence he objective audience, and increase client conversations.AI is he apparatus that controls conversational marketing

he client of today doesn't want to tend unfriendly calls, personal communication; they want swift and custom-made information. Voice facilitated conversational advertising because it can give direct responses and feedback

By using conversational marketing it can assist condense the sales sequence by providing instant and important responses to probable customers and reducing on the time required on making and succeeding leads.

Conversational marketing is altering the way business dialogue with their clients. It overlays the way for collaborative and modified communications .AI powered conversational marketing scrutinizes through the selling noise and provides important communication at the precise time.

SALES AND MARKETING FORCAST

Some of the major benefits of AI in digital marketing is its capability to foretell client activities and predict sales. Using prognostic analytics, AI puts together data mining, numbers and auction modelling to foretell future results for online businesses.

AI aids teams to close extra business with bigger distinguish ability into their determination and effectiveness. It gathers auction activity data to understand where auctions teams use their time then use that data to find stagnant contracts.AI provides timely warning signals about pacts that are slowing down then constructs an activity founded achievement roadmap for every single chance.

AI uses this entrée to data to provide AI compelled sales coaching to aid auction reps decrypt why contracts are wedged and what to do subsequently qualifies them to target themselves beside top performers, and recognize what brands top performers unlimited.

AI has established a comprehensive behavioral analytics result that defines which auction rep activities that are most probable to close a contract.

1. CRYSTAL IDENTIFIES

Crystal provides sales experts character profiles for everybody they come interact with. It plans profile material across the web, in gears like linked, sales strength and more. Crystal grants them access to character motivated email template centered on the recipient's character. This sort of custom-made communication is intended to resonate with receivers, extending the relationship amid the auction pro and client or prospect.

Crystal works in cooperation with web and desktop founded software solicitations, where it actually shines in email submissions.

Crystal can bid suggestions on the dialect to use as auctions pros type, suggests phrasing ideas and template directly in the tools, making it easy to use

Crystals AI gives character perceptions at the point of necessity, facilitating sales experts mend their communication with client's prospects and colleagues

2. TROOPS

Allocated as slack bot for auctions force, troops aid sales experts mechanize their work flow and line them up with well-designed groups like marketing and merchandise management. It assists to pull info from auctions force and makes it stress-free to publicize, and changing a slack channel.

From auto report generations that are available in slack to looking for sale force client info in a specific slack frequency, troops merge data for sales experts in an individual spot. Sales supervisors like troops for cool, no- coding necessary way they assimilate sales force in slack, forming new information, control panel and alarms that keep their crews up-to-date on what is going on in actual time.

Troop's aid sales experts uphold their sales duct more effortlessly as it spontaneously tracks sales force information and bestowing it to them in a laid-back manner. Several clicks and it organize the rest.

3. QUILL

From narrative science Quill is a regular language producer for creative organizations including sales crews. It evaluates online and digital information to categorize the facts, lyrics, and linguistics that are vital to your auctions organization. It yields content of your auctions organization that meets your corporate guidelines and style inclinations like character, style, structuring and the lyrics you use.

Quill intensifies the importance of the sales data establishments have already and generates modified sales writings, reports etc. That liberates your sales team's time.

4. CLARA LABS

Clara is computerized and spontaneous meeting scheduler. She uses dialect to reply to emails demands, is accessible twenty four hours in a week, and monitors up spontaneously with meeting parties.

Clara saves sales experts time and work, reserving meetings simply and without difficulty.

MARKETING FUNNELS AND LEAD GENERATIONS

AI is transforming lead generation and dialog. In the long run, there will be a great transformative impact on businesses and vocations.

AI structures are hugely centered on colossal data volumes analysis. Once they analyze data they now apply those insights to generate predictions. Also AI systems learn and progress over periods on their own as they are exposed to more data.

We could say that AI integration in marketing funnels and lead generations create platforms for organizational and careers growth at an impressive rate.

ACQUIRE VITAL INSIGHTS ABOUT LEADS

Huge data sets can now be analyzed in an instance thanks to AI technology and thus more accurate decisions and predictions can be reached at.

AI can analyze previous lead patterns than convey to you which leads to pay attention to.

BUILD YOUR LEAD DATABASE

Based on information about current leads, AI tools can find you more leads hence emphasizing on the importance of AI integration.

AI SOCIAL MEDIA ENGAGEMENT

Compared to various facets of people's lives, AI manipulates online media advertising in various dimensions. As a result there has been huge impact of Artificial Intelligence in the segment. Also a number of the top major leading major labels have joined onboard.

Because of cumulative request for public grid advertising which utilize smart systems. Numerous companies have upped their game to offer this provision. Currently several corporations are aiding business to grow by involving objective viewers in public media podiums.

APPLICATION OF PICTURE RECOGNITION IN ADVERTISING

Facing escalation of machine intelligence in public mass media marketing, pictures have a different resolve. Intelligent tech has aided public network dealers enabling efficient image usage to increase commitment rates. Additionally salespersons now have picture recognition software to find forms in customer activities. By utilizing AI enabled software merchandise and amenities agents aim at definite clients modifying content on their portals.

INTRODUCTION OF FACE IDENTIFICATION IN SOCIAL MEDIA MARKETING

Through machine intelligence public media salespersons can now incorporate facial recognition in public media marketing. Example of Facebook it uses face recognition to make the tagging purpose easy. Facebook recognizes the individual on the picture spontaneously saving the user the time to type the name of the person on the image.

Salespersons can benefit themselves from this feature by spontaneously tagging persons on the images of their merchandise, badges and so on.

UPRISING OF CHATBOTS AND VIRTUAL ASSISTANTS

Chatbots have aided marketers on public sites in various manners. This is by removing all presumptions from promotion. Formerly promoters had to chance when pushing their merchandise on public forums. This is because of inadequate information on what was efficient or not.

They have aided promoters to conduct in-depth scrutiny of different postings on public networks that are in line with their scope.

Chats have made it possible for promoters to forecast the likelihood of their posts doing well on various public forums.

IMPROVEMENT IN THE SOCIAL LISTENING PROCESS

Social listening can be defined as the procedure of tracing dialogues about certain words, slogans or trademarks and from the data collected to discover openings, compose information aiming a certain client segment on public media forums.

Using public location listening, salesperson can discover operational methods of connecting with clients in actual period and consequently study ways of molding promotion crusades to reach the customer more efficiently.

DIGITAL MARKETING TECHNOLOGIES

AI tools can aid generate content schemes that are resourceful supported by data. These schemes are

Hubspot

Hubspot utilizes machine learning to aid salespersons realize fresh content concepts that thrive and certify those content concepts in best proficient methods available.

Concured

It shows salespersons precisely what matters motivate commitment and what to scribble subsequently. The outcome is AI driven content scheme forums that mechanizes content review, subject search the generation of data based content guidelines content advertising and performance evaluation.

Crayon

Through machine learning to offer viable acumen on precisely what your rivals are applying online. You can monitor how the key folios of a corporation's website transform gradually.

Acronx

Acronx is a content approach placement forum that aid main enterprise trademarks better the value of their content at measure. By applying AI to go over content and ensure it suits your requirements regardless who the composer is.

Vennli

It based on content acumen forums that help marketers to be more appropriate to consumers in messaging, content strategy and marketing infrastructures.

Vennli grants the marketer the ability to gauge their content logistics over time, in order to measure performance at scale.

MARKETING FUNNELS AND LEAD GENERATIONS

AI is transforming lead generation and dialog. In the long run, here will be a great transformative impact on businesses and vocations.

AI structures are hugely centered on colossal data volumes analysis. Once they analyze data they now apply those insights to generate predictions. Also AI systems learn and progress over periods on their own as they are exposed to more data.

We could say that AI integration in marketing funnels and lead generations create platforms for organizational and careers growth at an impressive rate.

ACQUIRE VITAL INSIGHTS ABOUT LEADS

Huge data sets can now be analyzed in an instance thanks to AI technology and thus more accurate decisions and predictions can be reached at.

AI can analyze previous lead patterns then convey to you which leads to pay attention to.

BUILD YOUR LEAD DATABASE

Based on information about current leads, AI tools can find you more leads hence emphasizing on the importance of AI integration.

20 Tips to Use Artificial Intelligence immediately for Your Business

The trick in the current success in business today can be linked to the incorporation of Artificial Intelligence systems in the day to day running of the organization. Some of the tips in which you can make use of AI for business growth include:

- Use Analytics that are Artificial Intelligence based to enable you to make better decisions for your business
- Add automation in your structure to increase your marketing strategy which translates to increased sales
- Make use of supply chain management to help in better inventory management
- Use AI to improve the safety and maintenance of your business equipment
- Use AI systems in your hiring process; this will enable you to recruit the most qualified candidates for your organization
- Use AI to protect your business from cybercrimes and in fighting fraud; if not checked, these can lead to serious negative business impacts
- Incorporate AI systems in your business to help in expansion via self-driving technologies
- Use AI for customer support optimization using a chatbot
- Use AI in object, facial and image recognition
- Use AI to create smarter end products
- Do your research to find out the appropriate AI apps that will give you maximum benefits
- Use AI to deliver customer experiences that are personalized
- Use social media platforms that are AI powered as your marketing tool
- Use AI to increase manufacturing and output efficiency

- Use AI in optimizing business logistics
- Use AI in predicting performance
- Use AI in predicting behaviour
- Use AI in advertising and marketing your business
- Use AI in analyzing and managing your data
- Use AI in automating workloads

t is accurate to say that integration of Artificial Intelligence in marketing has greatly revolutionized marketing. We can only peculate and be optimistic on what the future holds for AI based marketing as the impact is already felt and we know that t can only get better as time and technology advances.

CONCLUSION

Artificial Intelligence is one major component in business in the modern world. Current trends indicate that AI is going to be even more incorporated in businesses in future. Important to note is that for the effective incorporation of AI into your businesses' analytics program, it's vital that strategies surrounding it goes hand in hand with the overall strategy of your business, always considering the merging of technology, people and processes. Artificial Intelligence is helping to input "increased smartness" to machines but it doesn't mean that it is necessarily ruling over the world.

It's very interesting to see the manner in which Artificial Intelligence is slowly making the running of businesses to be so much easier. The possibility of solving problems with close to zero error margins results in better running of your business as well as getting answers you require in even shorter periods. The main setback however is that in as much as this concept is very effective in running of businesses, it requires quite a huge capital for it to be installed. This basically mean that small-scale businesses may not find it cost effective to use AI in their companies, locking them out from enjoying AI benefits.

It can be concluded that the main aim of Artificial Intelligence in business is to facilitate the provision of software that has the ability to explain output and reason with connection to input.

AI provides interactions that are human-like as well as offering support in making decisions for certain tasks. However, Artificial Intelligence hasn't replaced human beings and it still won't, even in the days to come.

9 781801 111119